WITHDRAWN
UTSA LIBRARIES

COMPARATIVE EUROPEAN POLITICS

General Editors: Hans Daalder and Ken Newton

Editorial Board: Brian Barry, Franz Lehner, Arend Lijphart, Seymour Martin Lipset, Mogens Pedersen, Giovanni Sartori, Rei Shiratori, Vincent Wright

Political Data Handbook

COMPARATIVE EUROPEAN POLITICS

Comparative European Politics is a series for students and teachers of political science and related disciplines, published in association with the European Consortium for Political Research. Each volume will provide an up-to-date survey of the current state of knowledge and research on an issue of major significance in European government and politics.

OTHER TITLES IN THIS SERIES

Parties and Democracy: Coalition Formation and Government Functioning in Twenty States
Ian Budge and Hans Keman

Politics and Policy in the European Community (second edition)
Stephen George

Multiparty Government: The Politics of Coalition in Europe
Michael Laver and Norman Schofield

Self-Interest and Public Interest in Western Politics
Leif Lewin

Government and Politics in Western Europe: Britain, France, Italy, West Germany
Yves Mény

Localism and Centralism in Europe:
The Political and Legal Bases of Local Self-Government
Edward C. Page

Political Data Handbook OECD Countries

JAN-ERIK LANE, DAVID McKAY,
AND
KENNETH NEWTON

OXFORD UNIVERSITY PRESS
1991

Oxford University Press, Walton Street, Oxford OX2 6DP
Oxford New York Toronto
Delhi Bombay Calcutta Madras Karachi
Petaling Jaya Singapore Hong Kong Tokyo
Nairobi Dar es Salaam Cape Town
Melbourne Auckland
and associated companies in
Berlin Ibadan

Oxford is a trade mark of Oxford University Press

Published in the United States
by Oxford University Press, New York

© Jan-Erik Lane, David McKay, Kenneth Newton 1991

All rights reserved. No part of this publication may be reproduced,
stored in a retrieval system, or transmitted, in any form or by any means,
electronic, mechanical, photocopying, recording, or otherwise, without
the prior permission of Oxford University Press

British Library Cataloguing in Publication Data
Data available

Library of Congress Cataloging-in-Publication Data
Political Data handbook: OECD countries/edited by Jan-Erik Lane,
David Mckay, and Kenneth Newton.
p. cm. — (Comparative European politics)
Includes bibliographical references.
1. Statistics. 2. Political statistics. I. Lane, Jan-Erik.
II. Mckay, David H. III. Newton, Kenneth, 1940– . IV. Series.
HA155.P65 1991   320'.021—dc20   91–4022
ISBN 0–19–827718–0

Typeset by Pure Tech Corporation, Pondicherry, India
Printed and bound in
Great Britain by Bookcraft (Bath) Ltd.
Midsomer Norton, Avon

Acknowledgements

We owe a debt of thanks to a large number of colleagues who freely gave of their time to review and comment upon individual country data. The individuals involved were Francis Castles, Margaret Clark, Eric Damgaard, Clement H. Dodd, Giorgio Freddi, André Frognier, Hannes Gissurarson, W. K. Jackson, Efthalia Kalogeropoulou, Michael Laver, Franz Lehner, Leif Lewin, Ian Neary, Richard Simeon, Ben Soetendorp, Maria José Stock, Josep Vallès, Matti Wiberg, Colette Ysmal. While acknowledging the assistance of these readers we do, of course, remain responsible for any errors or misinterpretations. Above all we would like to thank Svante Errsson, without whom the book could never have been completed.

Contents

Introduction

PART I: COMPARATIVE TABLES

1. Population
   Population and population growth. Area and population density. Age structure. Birth rate, life expectancy, and infant mortality. Urban concentrations.   7

2. Social Structure
   Ethno-linguistic and religious structures. Income distribution. Unionization. Migration.   18

3. Employment
   Labour force and employment statistics. Public sector employment. Unemployment. Industrial disputes.   31

4. Economy
   GDP per capita. Origin of GDP by economic sector. Inflation. External trade dependency. Growth of GDP.   47

5. Public Finance
   General government: current receipts and disbursements, taxes, social security contributions and transfers, final consumption, service expenditures, military expenditure, development assistance, deficits.
   Central government: current receipts and disbursements; taxes, final consumption, transfers, deficits, expenditures by function.
   State and local government: taxes and final consumption.   64

6. Government Structures
   Chamber systems. Electoral systems. Electorates and suffrage. Constitutional developments. State, regional, and local government. Government formation in Western Europe.   107

7. Political Parties and Elections
National electoral participation. National political parties. Electoral strength of communist, religious, socialist, ethnic, agrarian, left/socialist, liberal, and conservative parties.   122

PART II: COUNTRY TABLES

1.	Austria	151
2.	Belgium	155
3.	Denmark	160
4.	Finland	164
5.	France	170
6.	Federal Republic of Germany	178
7.	Greece	182
8.	Iceland	186
9.	Ireland	189
10.	Italy	192
11.	Luxembourg	198
12.	The Netherlands	201
13.	Norway	205
14.	Portugal	209
15.	Spain	213
16.	Sweden	216
17.	Switzerland	220
18.	Turkey	224
19.	United Kingdom	227
20.	Canada	231
21.	USA	234
22.	Japan	239
23.	Australia	243
24.	New Zealand	247

Data Archives for the Social Sciences   251

References   255

Introduction

The purpose of this book is to provide, within the confines of a single and reasonably priced volume, as comprehensive and detailed a statistical guide as possible to the government and politics of the twenty-four OECD countries, together with essential social and economic background information. The aim is to present essential material of a factual nature in a form which meets the needs of professional politicians, administrators, journalists, teachers, political scientists, and the interested citizen.

There is, of course, no lack of statistical data about the nations which form the OECD—the Organization for Economic Co-operation and Development. They are, by and large, the most advanced of modern societies and they generate singly and collectively a huge amount of information about themselves, but most of it is not easily available. For example, they have their own official statistical yearbooks, but complete sets of these are held by only a few of the largest and most specialized libraries. Also, various international organizations produce large amounts of statistical material dealing with such things as social trends, domestic and international trade and business, banking, public expenditure, labour markets, migration, and so on, but these figures are often difficult to understand and interpret, often contained in voluminous and very expensive publications, and rarely deal with government and politics. There are also a handful of more specialized political source books, but these tend to be expensive and sometimes difficult to find.

Those interested, therefore, usually have to plough through a fairly long list of different sources to get even the most elementary data, and even then they are likely to end up with gaps, inconsistencies, and inadequacies. To complete this book we had to comb very carefully through nearly 230 sources, and sift through 450 tables—sometimes three or four or five dealing with almost identical variables—before we could find a suitable one. Some of our sources are well known, even if they are not always readily available. Others are relatively obscure, usually costly, and invariably not held by local libraries.

Our original intention was to compile a handbook for Western Europe, because it is so sorely needed. Western Europe shows such a richness and variety of conditions and circumstances that it makes a marvellous natural laboratory for social scientists—yet it has no political handbook. But it is also the case that the major nations of Western Europe include most of the democracies of the world, and all but a few of its wealthiest and most advanced urban-industrial nations. It takes little extra effort to cover the comparable countries outside Western Europe, and it makes the volume even more useful as a source book to do so. Therefore, it was decided to cover not just the nineteen countries of Western Europe, but also the remaining five advanced democracies of the West—namely, Canada, the USA, Japan, Australia, and New Zealand. We therefore cover twenty-four nations in all, but format the statistical tables in such a way as to distinguish between the nineteen European countries, and the 'rest'. We believe that the countries are sufficiently alike to make comparison of them meaningful, and, at the same time, they provide a sufficiently large number of cases to make many forms of statistical analysis worthwhile.

The book is divided into two main sections. The first consists of comparative tables covering all countries for as much of the post-World War II period as possible. It is divided, in turn, into sections covering the following sorts of data: population structure, employment, economy, industry, public expenditure and taxation, government structure, and electoral data. Each table has notes which draw the reader's attention to the sources used, to definitions, and to various complications and observations about the figures. In the majority of cases the time period covered is 1950 to 1985, usually for five- or ten-year intervals, depending upon the nature and availability of the material to hand.

The second main part of the book consists of sections which cover the most significant features of government and politics in the twenty-four nations which cannot be easily produced in tabular form. Each section provides information about state structure and offices, parties, government, constitutions and their changes, electoral and voting systems, as well as basic material about economic interest organizations and the media.

We cannot hope to present everything that everybody ever wanted to know about the politics and society of twenty-four nations over a thirty-five-year period. Consequently we have added a third, shorter section which lists sources of further information, including both the sources which we have consulted and quoted, and other published work, and Western European data archives which typically carry an extensive array of machine readable data files covering many different aspects of government, politics, and society in Western Europe.

Although the few paragraphs above briefly outline the purposes and contents of the book, they say nothing about why it comes to have its particular contents, and something by way of general explanation should be provided on this aspect of the volume. It is both paradoxical and ironic that something as objective and statistical as a data handbook should require so many subjective judgements to be made in the process of compilation. In fact, at almost every turn we had to make difficult decisions about what to include, why, and how. There was the first intention to produce a book about Western Europe, subsequently modified to include all twenty-four OECD countries, for reasons just explained. There was then the decision about the time period covered. In theory it would be ideal to cover the twentieth century, but this is not practicable, for reasons of space, and because the further one goes back in time, the more difficult it is to find data, never mind valid and reliable data. If one then settles for the period after World War II, when good data increasingly becomes available, it would be best to include everything from 1945 to the latest year. However, information before 1950 is often thin and scattered. At the other end of the period, it usually takes some time for statistical material to be produced, so that any book such as this one is bound to be a couple of years or more out of date. (As we go to press, West and East Germany have just been unified.) Consequently, we decided to start the statistical runs in 1950, and to end with the the most recent available information, although this is often for 1985 or earlier.

Hard decisions then had to be made about what variables to cover, and which set of statistics on a given variable to include. The general policy has been to cover only basic background information on social and economic matters, so as to leave as much room as possible for political and governmental ones. There are, after all, quite a few sources of social and economic statistics, but few on political and governmental ones, and it is this deficiency which we wanted to remedy most urgently.

On some variables it is possible to assemble as many as four or five tables from different sources, and yet they rarely manage to agree entirely. It must be said, however, that the discrepancies are usually negligible, amounting to little more than rounding errors, and in these cases we opted for the set of figures which was most complete or which covered the longest time span, or both. Where the discrepancies were large, and on rare occasions they were substantial, we inspected the footnotes with extra care in order to select the figures which made most sense for this volume, and which were most compatible with other tables in it.

Finally, we had to make a basic policy decision about exactly what sort of statistics should be included: should it be basic data of a fairly 'raw' kind; should we process this data so as to produce various indices of our own design which would allow the reader to make direct comparisons between nations over a wide range of different sorts of indicators; or should we include the indices of other writers and authors who have developed them for their own particular purposes? We decided against the second on the grounds that the sort of figures which people want for their own specific purposes are usually so diverse and precise that it is best to give them the basic ingredients, with which they can work as they wish. Nevertheless, we have not provided completely raw data, but have standardized it on a 'per capita' or 'as a percentage of GDP' basis, as appropriate. This means that the tables make sense in their own right, and that the statistics can be used and manipulated in various ways for more elaborate analysis. At the same time, the social science literature does contain some indices of such general interest and importance that we have included them here. Throughout the volume we have used two abbreviations in the following way: a dash, '—', stands for information not available whereas 'n.a.' means not applicable.

We hope the resulting compilation is as useful as it is fascinating, whether it is used casually to check up on the odd fact, or for a more systematic comparison of the twenty-four most advanced political systems of the Western world. We all know that there are 'lies, damn lies, and statistics', and we all know that statistics do not speak for themselves. But if those who speak for the statistics in political life are not to be damn liars, then they—and we—must have a supply of reliable and valid data to hand. We hope that this volume provides just such a collection.

Jan-Erik Lane
David McKay
Kenneth Newton

PART I
Comparative Tables

Section 1: Population

TABLES

1.1. Population: mid-year estimates in thousands
1.2. Population growth: average annual percentage changes
1.3. Area in km^2, population density, and inhabitants per km^2
1.4. Age structure: population aged 0 to 14 as a percentage of total population
1.5. Age structure: population aged 15 to 64 as a percentage of total population
1.6. Age structure: population aged 65 and older as a percentage of total population
1.7. Crude birth rate (per thousand)
1.8. Life expectancy at birth in years
1.9. Infant mortality rates (up to one year)
1.10. Urban concentration: largest city's percentage share of total population

Comparative Tables

TABLE 1.1. Population: mid-year estimates in thousands

	1950	1960	1970	1980	1985
Aus.	6 935	7 048	7 426	7 546	7 545
Belg.	8 639	9 153	9 656	9 859	9 853
Den.	4 271	4 581	4 951	5 123	5 101
Fin.	4 009	4 430	4 606	4 779	4 919
Fr.	41 736	45 684	50 768	53 713	55 133
FRG	49 986	55 423	60 714	61 561	61 065
Gr.	7 566	8 327	8 793	9 642	9 937
Ice.	143	176	204	228	241
Ire.	2 969	2 832	2 944	3 307	3 560
It.	47 104	50 198	53 661	56 159	56 945
Lux.	296	314	339	363	366
Neth.	10 114	11 480	13 032	14 144	14 486
Norw.	3 265	3 581	3 877	4 086	4 144
Port.	8 405	8 826	9 044	9 752	10 198
Sp.	28 009	30 455	33 779	37 430	38 730
Sw.	7 047	7 480	8 043	8 311	8 330
Switz.	4 694	5 362	6 267	6 349	6 421
Turk.	20 809	27 509	35 321	44 438	49 406
UK	50 616	52 372	55 416	55 944	56 539
Can.	13 737	17 909	21 324	23 941	25 414
USA	152 271	180 671	205 052	227 658	238 780
Jap.	82 900	94 094	104 345	116 782	120 579
Austral.	8 179	10 275	12 507	14 616	15 789
NZ	1 908	2 372	2 820	3 268	3 246

Sources: 1950–80: World Bank (1984): vol. i; 1985: World Bank *The World Bank Atlas* (1987).

TABLE 1.2. Population growth: average annual percentage changes

	1950–60	1960–70	1970–80	1980–5	1950–85
Aus.	0.15	0.49	0.15	0.00	0.24
Belg.	0.54	0.50	0.19	−0.01	0.39
Den.	0.66	0.78	0.32	−0.07	0.54
Fin.	0.95	0.36	0.34	0.49	0.63
Fr.	0.86	1.01	0.59	0.44	0.89
FRG	0.99	0.87	0.19	−0.19	0.62
Gr.	0.91	0.51	0.88	0.51	0.87
Ice.	2.10	1.45	1.07	0.95	1.90
Ire.	−0.42	0.36	1.12	1.28	0.55
It	0.60	0.63	0.42	0.23	0.56
Lux.	0.55	0.72	0.64	0.14	0.66
Neth.	1.23	1.23	0.78	0.40	1.20
Norw.	0.66	0.75	0.49	0.24	0.75
Port.	0.46	0.22	0.71	0.76	0.59
Sp.	0.79	0.99	0.98	0.58	1.06
Sw.	0.56	0.68	0.30	0.04	0.51
Switz.	1.29	1.59	0.12	0.19	1.02
Turk.	2.93	2.58	2.35	1.86	3.82
UK	0.32	0.59	0.09	0.18	0.33
Can.	2.76	1.79	1.12	1.03	2.36
USA	1.70	1.29	1.00	0.81	1.58
Jap.	1.29	0.99	1.08	0.54	1.26
Austral.	2.33	1.97	1.53	1.34	2.58
NZ	2.21	1.72	1.44	−0.11	1.95

Note: The data for 1950–60 are averaged over the eleven year period; data for subsequent periods are average annual percentage changes, i.e. the figures for 1950–60 are arrived at in the following way: ((Pop. 1960 − Pop. 1950) × 100)/(Pop. 1950 × 11).

Sources: See Table 1.1

TABLE 1.3. Area in km^2, population density, and inhabitants per km^2

	Area		Density	
	1955	1980	1955	1980
Aus.	83 849	83 849	83	90
Belg.	30 507	30 513	283	323
Den.	43 042	43 069	99	119
Fin.	337 009	337 032	12	14
Fr.	551 208	547 026	76	98
FRG	248 407	248 577	201	248
Gr.	132 526	131 944	57	73
Ice.	103 000	103 000	1	2
Ire.	70 283	70 283	42	47
It.	301 226	301 225	156	186
Lux.	2 586	2 586	114	140
Neth.	32 450	40 844	312	346
Norw.	323 917	324 219	10	13
Port.	92 200	92 082	91	106
Sp.	503 486	504 782	56	74
Sw.	449 682	449 964	16	18
Switz.	41 288	41 288	114	154
Turk.	776 980	780 576	27	57
UK	244 016	244 046	207	229
Can.	9 974 375	9 976 139	1	2
USA	9 357 427	9 372 614	16	24
Jap.	369 766	377 708	224	309
Austral.	7 704 159	7 686 848	1	2
NZ	267 995	268 676	7	12

Sources: 1955: UN, *Demographic Yearbook* (1957); 1980: UN, *Demographic Yearbook* (1981).

TABLE 1.4. Age structure: population aged 0 to 14 as a percentage of total population

	1950[a]	1960	1970	1980	1985
Aus.	22.9	22.1	24.4	20.4	18.2
Belg.	20.6	23.5	23.6	20.0	18.7
Den.	26.3	25.2	23.3	20.9	18.4
Fin.	30.0	30.4	24.6	20.3	19.4
Fr.	22.6	26.4	24.8	22.4	21.2
FRG	23.3	21.3	23.2	18.2	15.1
Gr.	28.4	26.5	24.6	22.8	20.9
Ice.	30.7	34.7	32.6	27.5	26.2
Ire.	28.9	31.1	31.3	30.4	29.2
It.	26.1	24.9	22.9	20.5	17.6
Lux.	19.8	21.3	22.1	18.6	17.1
Neth.	29.3	30.0	27.3	22.3	19.5
Norw.	24.4	25.9	24.5	22.2	20.0
Port.	29.4	29.2	28.8	25.3	23.1
Sp.	26.2	27.4	28.0	25.9	23.1
Sw.	23.4	22.0	20.9	19.6	18.2
Switz.	23.6	23.6	23.4	19.5	17.5
Turk.	38.3	41.2	41.8	38.5	36.6
UK	22.6	23.3	24.1	21.0	19.2
Can.	30.3	33.5	30.3	23.0	21.5
USA	25.9	31.0	28.3	22.5	21.7
Jap.	35.4	30.2	23.9	23.6	21.6
Austral.	25.2	30.1	28.8	25.3	23.6
NZ	29.4	32.9	31.9	27.2	24.6

[a] 1950 or nearest available year.

Note: Country totals for Tables 1.4 to 1.6 do not always add up to 100% because of rounding.

Sources: 1950: Mitchell (1981; 1982; 1983); UN, *Demographic Yearbook* (1949–50; 1952); 1960: World Bank (1984): vol. ii; 1970–85: OECD (1988).

TABLE 1.5. Age structure: population aged 15 to 64 as a percentage of total population

	1950	1960	1970	1980	1985
Aus.	66.5	65.8	61.6	64.2	67.5
Belg.	68.8	64.5	63.0	65.6	67.4
Den.	64.6	64.2	64.4	64.7	66.5
Fin.	63.4	62.4	66.2	67.7	68.1
Fr.	66.1	62.0	62.3	63.7	65.8
FRG	67.3	67.8	63.6	66.3	70.0
Gr.	65.0	65.3	64.3	64.0	65.7
Ice.	61.6	57.4	58.5	62.6	63.7
Ire.	60.4	57.7	57.7	58.8	60.0
It.	65.7	65.8	66.5	66.7	69.5
Lux.	70.7	67.8	65.3	67.8	69.7
Neth.	63.6	61.0	62.6	66.2	68.5
Norw.	65.9	63.0	62.6	63.1	64.3
Port.	63.3	62.9	61.9	63.3	64.9
Sp.	66.5	64.4	62.5	63.3	64.9
Sw.	66.2	66.0	65.5	64.1	64.6
Switz.	66.8	66.3	65.2	66.8	68.6
Turk.	58.4	55.2	53.8	56.6	59.3
UK	66.6	65.1	62.7	64.1	65.6
Can.	61.9	59.0	61.7	67.5	68.1
USA	62.6	59.7	61.9	66.2	66.4
Jap.	58.7	64.1	69.0	67.4	68.2
Austral.	66.8	61.4	62.8	65.1	66.2
NZ	61.3	58.5	59.6	63.1	65.1

Sources and notes: See Table 1.4.

TABLE 1.6. Age structure: population aged 65 and older as a percentage of total population

	1950	1960	1970	1980	1985
Aus.	10.6	12.0	14.1	15.4	14.3
Belg.	10.6	12.0	13.4	14.4	13.8
Den.	9.1	10.6	12.3	14.4	15.1
Fin.	6.6	7.2	9.1	12.0	12.5
Fr.	11.2	11.6	12.9	13.9	12.9
FRG	9.4	10.8	13.2	15.5	14.8
Gr.	6.6	8.3	11.1	13.1	13.4
Ice.	7.7	8.0	8.9	9.9	10.1
Ire.	10.7	11.2	11.2	10.7	10.8
It.	8.2	9.3	10.5	12.9	12.9
Lux.	9.5	10.8	12.6	13.6	13.2
Neth.	7.1	9.0	10.2	11.5	12.1
Norw.	9.7	11.1	12.9	14.8	15.7
Port.	7.3	8.0	9.3	11.5	12.0
Sp.	7.3	8.2	9.5	10.9	12.0
Sw.	10.3	12.0	13.7	16.3	17.2
Switz.	9.6	10.1	11.4	13.7	13.9
Turk.	3.3	3.5	4.4	4.8	4.1
UK	10.9	11.7	13.2	14.9	15.1
Can.	7.8	7.5	8.0	9.5	10.4
USA	11.5	9.2	9.8	11.3	11.9
Jap.	4.9	5.7	7.1	9.0	10.2
Austral.	8.0	8.5	8.3	9.6	10.2
NZ	9.2	8.6	8.4	9.7	10.2

Sources and notes: See Table 1.4.

TABLE 1.7. Crude birth rate (per thousand)

	1950	1960	1970	1980	1985
Aus.	15.6	17.9	15.1	12.1	12.2
Belg.	16.9	16.9	14.7	12.6	11.9
Den.	18.6	16.6	14.3	11.2	10.4
Fin.	24.5	18.5	14.0	13.1	13.2
Fr.	20.6	17.9	16.8	14.9	14.0
FRG	16.2	17.5	13.4	10.1	9.7
Gr.	20.2	18.9	16.5	15.4	13.0
Ice.	28.7	17.5	13.4	19.9	18.3
Ire.	21.3	21.4	21.9	21.8	19.0
It.	19.6	18.1	16.8	11.4	10.0
Lux.	14.8	16.0	13.2	11.4	11.4
Neth.	22.7	20.8	18.3	12.8	11.9
Norw.	19.1	17.3	16.6	12.4	12.2
Port.	24.4	24.2	19.1	16.2	14.0
Sp.	20.2	21.7	19.4	15.1	13.0
Sw.	16.4	13.7	13.7	11.6	11.3
Switz.	18.1	17.6	15.8	11.5	11.4
Turk.	—	43.1	37.9	31.4	30.1
UK	16.2	17.5	16.3	13.5	13.0
Can.	27.1	26.7	17.4	15.4	14.7
USA	23.5	23.6	18.2	15.9	15.0
Jap.	28.2	17.3	18.7	13.5	12.5
Austral.	23.3	22.4	20.6	15.3	15.0
NZ	25.9	26.5	22.1	16.2	16.0

Note: Crude birth rate equals annual live births per thousand population.

Sources: 1950: UN, Demographic Yearbook (1954); 1960–70: World Bank (1984): vol. ii; 1980–5: World Bank (1988).

TABLE 1.8. Life expectancy at birth in years

	1960	1970	1980	1985
Aus.	68.6	70.5	72.5	73.6
Belg.	70.3	71.9	72.8	75.0
Den.	72.2	73.3	74.1	75.4
Fin.	68.4	70.0	72.9	75.5
Fr.	70.3	72.1	74.3	78.3
FRG	69.5	70.5	73.2	75.0
Gr.	68.8	71.3	74.2	75.0
Ice.	69.5	70.5	76.5	77.0
Ire.	69.6	71.4	72.5	73.6
It.	69.2	71.5	74.5	76.8
Lux.	68.1	70.4	72.4	73.5
Neth.	73.2	73.7	75.7	76.6
Norw.	73.4	74.2	76.1	77.0
Port.	63.3	67.1	70.1	73.8
Sp.	68.5	71.5	74.5	77.0
Sw.	73.1	74.3	76.3	77.1
Switz.	71.2	72.3	77.0	76.8
Turk.	50.5	56.6	61.7	64.1
UK	70.6	71.8	73.8	74.6
Can.	71.0	72.2	74.6	76.0
USA	69.9	70.9	73.7	75.8
Jap.	68.0	72.4	76.0	77.1
Austral.	70.9	72.1	74.6	77.5
NZ	71.7	71.8	73.3	74.3

Note: Life expectancy at birth is the number of years a newborn infant would live if prevailing patterns of mortality for all people at the time of birth were to stay the same throughout life.

Sources: 1960–70: World Bank (1984): vol. ii; 1980–5: World Bank (1988).

Comparative Tables

TABLE 1.9. Infant mortality rates (up to one year)

	1950	1960	1970	1980	1985
Aus.	66.1	37.5	25.9	13.9	11.0
Belg.	53.4	31.2	21.1	11.0	9.4
Den.	30.7	21.5	14.2	8.4	7.0
Fin.	43.5	21.0	13.2	7.6	6.0
Fr.	52.0	27.4	18.2	10.0	8.0
FRG	55.5	33.8	23.6	12.7	10.0
Gr.	36.7	40.1	29.6	17.9	16.0
Ice.	21.7	15.2	13.2	11.1	10.1
Ire.	45.3	29.3	19.5	11.1	10.0
It.	63.8	43.9	29.6	14.3	12.0
Lux.	45.7	31.5	24.6	11.5	11.7
Neth.	25.2	17.9	12.7	8.6	8.0
Norw.	28.2	18.9	12.7	8.0	8.0
Port.	94.1	82.0	58.0	23.9	19.0
Sp.	69.8	49.6	26.5	11.1	10.0
Sw.	21.0	16.6	11.0	6.9	6.0
Switz.	31.2	21.1	15.1	8.4	8.0
Turk.	—	189.5	147.5	122.6	92.1
UK	31.4	22.5	18.4	12.1	9.0
Can.	41.3	27.3	18.8	10.4	8.0
USA	29.2	26.0	20.0	12.6	11.0
Jap.	60.1	30.4	13.1	7.5	6.0
Austral.	24.5	20.2	17.9	10.7	9.0
NZ	27.6	22.6	16.7	12.8	11.0

Note: Infant mortality rate is the number of infants who die before reaching one year of age, per thousand live births in a year.

Sources: 1950 UN, *Demographic Yearbook* (1954); 1960–70: World Bank (1984): vol. ii; 1980–5: World Bank (1988).

TABLE 1.10. Urban concentration: largest city's percentage share of total population

		1950	1960	1970	1980
Aus.	(Wien)	25	23	22	21
Belg.	(Bruxelles)	11	11	11	10
Den.	(København)	27	28	28	27
Fin.	(Helsinki)	9	10	11	10
Fr.	(Paris)	7	6	5	4
FRG	(Berlin West)	4	4	4	3
Gr.	(Athínai)	7	8	10	9
Ice.	(Reykjavik)	43	41	40	38
Ire.	(Dublin)	18	19	19	16
It.	(Roma)	4	4	5	5
Lux.	(Luxembourg)	21	23	23	22
Neth.	(Amsterdam)	8	8	6	5
Norw.	(Oslo)	13	13	13	11
Port.	(Lisboa)	9	9	8	8
Sp.	(Madrid)	6	7	9	9
Sw.	(Stockholm)	11	11	10	8
Switz.	(Zürich)	8	8	7	6
Turk.	(Istanbul)	5	5	6	6
UK	(London)	17	16	13	12
Can.	(Toronto)	5	4	3	3
USA	(New York)	5	4	4	3
Jap.	(Tokyo)	8	9	8	7
Austral.	(Sydney)	18	21	20	22
NZ	(Auckland)	17	17	11	10

Note: City refers to proper city as defined in the sources. There is a problem of comparability both between nations and over time, since proper city may be defined in various ways. The UN's *Demographic Yearbook* states that *city proper* is defined as a locality with legally fixed boundaries and an administratively recognized urban status which is usually characterized by some form of local government.

Sources: Figures for largest city 1950–70 are based on: Mitchell (1981; 1982; 1983); figures for 1980 are based on UN, *Demographic Yearbook* (1983).

Section 2: Social Structure

TABLES

2.1. Ethno-linguistic structure: percentage of the population belonging to dominant groups
2.2. Ethno-linguistic structure: fragmentation indices
2.3. Religious structure: religious affiliation by Christian denomination, other denominations, or no denomination in the mid-1970s, (percentages)
2.4. Income distribution: Gini-indices
2.5. Income distribution: share of national income (pre- or post-tax) going to top 20% of the population
2.6. Estimates of percentage of work-force unionized 1950–1980
2.7. Estimates of work-force unionized (percentages)
2.8. Migration
2.9. Migration in Western Europe: minority labour force as percentage of total labour force
2.10. Migration in Western Europe: minority population as percentage of total population
2.11. Migration in Western Europe: estimated population and labour force living abroad

TABLE 2.1. Ethno-linguistic structure: percentage of the population belonging to dominant groups

	Language group (1)	Ethno-linguistic group (2)
Aus.	99	93
Belg.	50	59
Den.	97	98
Fin.	92	92
Fr.	86	82
FRG	100	94
Gr.	98	95
Ice.	97	99
Ire	97	96
It.	99	96
Lux.	93	82
Neth.	95	94
Norw.	100	98
Port.	100	99
Sp.	73	73
Sw.	98	90
Switz.	69	63
Turk.	90	90
UK	98	79
Can.	58	38
USA	86	63
Jap.	99	98
Austral.	91	83
NZ	91	89

Sources: (1): Data are for the 1950s and reported in Rustow (1967); (2): Data are for the 1970s and reported in Barrett (1982).

TABLE 2.2. Ethno-linguistic structure: fragmentation indices

	(1)	(2)	(3)	(4)	(5)	(6)
Aus.	.04	.01	.13	.02	.02	.13
Belg.	.55	.51	.55	.50	.50	.54
Den.	.02	.12	.05	.00	.02	.05
Fin.	.16	.35	.16	.13	.11	.15
Fr.	.15	.17	.26	.10	.15	.33
FRG	.00	—	.03	.00	.02	.12
Gr.	.03	.21	.10	.09	.04	.10
Ice.	.00	.54	.05	.00	.00	.03
Ire.	.22	.44	.04	.04	.18	.07
It.	.06	.09	.04	.10	.10	.08
Lux.	.00	.12	.15	.00	.14	.32
Neth.	.08	.00	.10	.06	.04	.11
Norw.	.00	.19	.04	.00	.02	.05
Port.	.00	.00	.01	.00	.00	.01
Sp.	.50	.42	.44	.39	.63	.44
Sw.	.00	.12	.08	.00	.02	.19
Switz.	.45	.50	.50	.52	.53	.56
Turk.	—	.24	.25	—	.19	.19
UK	.05	.04	.32	.03	.02	.37
Can.	—	.48	.75	—	.56	.77
USA	—	.25	.50	—	.31	.58
Jap.	—	.02	.01	—	.00	.03
Austral.	—	.03	.32	—	.02	.32
NZ	—	.07	.37	—	.27	.20

Note: The index of fragmentation indicates the probabilities that two randomly sampled people will belong to different ethno-linguistic groups. The index ranges from 0 to 1, with 1 the most heterogeneous, and 0 the least.

Sources: (1): Data are for the 1920s and based on Tesnière (1928); (2): Taylor and Hudson (1972): Muller data; (3): Taylor and Hudson (1972): Atlas Narodov Mira data; (4): Data are for the 1970s and based on Stephens (1976); (5): Data are for the 1970s and based on Worldmark (1984); (6): Data are for the 1970s and based on Barrett (1982).

TABLE 2.3. Religious structure: religious affiliation by Christian denomination, other denominations, or no denomination in the mid-1970s, (percentages)

	Protestants	Catholics	Other religions	No religion/ non-confession	Fragmentation index
Aus.	7	90	0	3	.18
Belg.	0	92	1	7	.15
Den.	95	1	1	3	.10
Fin.	94	1	0	5	.11
Fr.	2	80	4	14	.34
FRG	48	46	2	4	.56
Gr.	0	98	2	0	.04
Ice.	97	1	0	2	.06
Ire.	4	96	0	0	.08
It.	0	87	0	13	.23
Lux.	1	94	0	5	.11
Neth.	43	44	2	11	.61
Norw.	98	0	0	2	.04
Port.	1	96	0	3	.08
Sp.	0	97	0	3	.06
Sw.	71	2	0	27	.42
Switz.	46	52	1	1	.52
Turk.	0	1	99	0	.02
UK	74	14	3	9	.42
Can.	43	50	2	5	.56
USA	58	31	5	6	.56
Jap.	2	1	87	11	.25
Austral.	57	32	1	10	.56
NZ	75	18	1	6	.40

Note: The fragmentation index refers to the probability that two randomly sampled people will belong to different religions. The index ranges from 0 to 1, with 0 the most homogeneous and 1 the most heterogeneous.

Sources: Barrett (1982).

TABLE 2.4. Income distribution: Gini-indices

	Index (1)	(Yr)	Index (2)	(Yr)	Index (3)	(Yr)	Index (4)	(Yr)	Index (5)	(Yr)
Aus.	—		0.369	(67)	—		—		—	
Belg.	—		—		—		—		—	
Den.	0.37	(63)	0.365	(66)	—		0.38	(71)	—	
Fin.	0.46	(62)	0.463	(67)	—		0.37	(71)	0.30[a]	(78)
Fr.	0.50	(62)	0.421	(70)	0.416	(70)	0.414	(70)	0.43	(75)
FRG	0.45	(64)	0.392	(70)	0.396	(73)	0.383	(73)	0.31[a]	(78)
Gr.	0.38	(57)	0.394	(57)	—		—		—	
Ice.	—		—		—		—		—	
Ire.	—		—		—		—		—	
It.	0.40	(48)	0.397	(69)	—		0.398	(69)	0.37	(80)
Lux.	—		—		—		—		—	
Neth.	0.42	(62)	0.393	(67)	0.385	(67)	0.354	(67)	0.31[a]	(77)
Norw.	0.35	(63)	0.361	(70)	0.354	(70)	0.307	(70)	0.32[a]	(79)
Port.	—		—		—		—		—	
Sp.	—		0.391	(64)	—		0.355	(73)	0.39	(80)
Sw.	0.39	(63)	0.350	(72)	0.346	(72)	0.302	(72)	0.33[a]	(79)
Switz.	—		0.401	(68)	—		—		0.29	(80)
Turk.	—		0.553	(68)	—		—		—	
UK	0.38	(64)	0.346	(62)	0.344	(73)	0.318	(73)	0.33	(80)
Can.	—		0.383	(69)	0.382	(69)	0.354	(69)	0.36	(79)
USA	0.34	(69)	0.409	(72)	0.404	(72)	0.381	(72)	—	
Jap.	0.39	(69)	0.412	(71)	0.335	(69)	0.316	(69)	0.17	(79)
Austral.	0.32	(66)	0.320	(68)	0.313	(67)	0.312	(67)	0.36	(79)
NZ	—		0.353	(72)	—		—		—	

[a] Figures based on authors' computations.

Note: The Gini index ranges from 0 to 1, with 0 being maximum equality and 1 maximum inequality. See Sawyer (1976) and OECD (1986) for a discussion of the methodological problems.

Sources: (1) Paukert (1973): pre-tax; (2) Bornischer (1978): pre-tax; (3) Sawyer (1976): pre-tax; (4) Sawyer (1976): post-tax; Denmark, Finland: Uusitalo (1975); (5) OECD (1986).

TABLE 2.5. Income distribution: share of national income (pre- or post-tax) going to top 20% of the population

	Share (Yr) (1)	Share (Yr) (2)	Share (Yr) (3)	Share (Yr) (4)	Share (Yr) (5)	Share (Yr) (6)
Aus.	—	—	—	—	—	—
Belg.	—	—	—	—	—	—
Den.	43.2 (63)	47.6 (68)	—	—	36.0 (78)	—
Fin.	49.3 (62)	49.3 (62)	—	—	38.6 (81)	—
Fr.	53.7 (62)	53.7 (62)	46.9 (70)	46.9 (70)	37.6 (81)	37 (78)
FRG	45.6 (70)	52.9 (64)	45.6 (70)	46.2 (73)	45.8 (75)	47 (75)
Gr.	—	49.5 (57)	—	—	39.5 (78)	38 (78)
Ice.	—	—	—	—	—	—
Ire.	—	—	—	—	39.4 (73)	—
It.	—	48.4 (48)	46.5 (70)	46.5 (69)	43.9 (77)	44 (80)
Lux.	—	—	—	—	—	—
Neth.	40.5 (63)	48.5 (67)	42.9 (70)	42.9 (67)	36.2 (81)	37 (77)
Norw.	—	40.5 (68)	37.3 (70)	37.3 (70)	38.2 (82)	37 (79)
Port.	—	—	—	—	49.1 (73)	—
Sp.	45.2 (64)	45.7 (65)	42.2 (70)	42.2 (74)	40.0 (80)	42 (80)
Sw.	42.5 (70)	44.0 (63)	37.0 (70)	37.0 (72)	41.7 (81)	37 (79)
Switz.	—	—	45.9 (70)	—	38.0 (78)	37 (78)
Turk.	60.6 (68)	60.8 (68)	56.5 (73)	56.5 (73)	56.5 (73)	—
UK	39.2 (68)	39.0 (68)	39.4 (70)	38.8 (73)	39.7 (79)	39 (80)

TABLE 2.5 Continued

	Share (Yr) (1)	Share (Yr) (2)	Share (Yr) (3)	Share (Yr) (4)	Share (Yr) (5)	Share (Yr) (6)
Can.	40.2 (65)	40.2 (65)	41.0 (69)	41.0 (69)	40.0 (81)	40 (79)
USA	38.8 (70)	38.8 (70)	42.8 (72)	42.8 (72)	39.9 (80)	—
Jap.	43.8 (68)	40.0 (63)	41.0 (69)	41.0 (69)	37.5 (79)	30 (79)
Austral.	38.7 (68)	38.8 (68)	38.8 (67)	38.8 (67)	47.1 (76)	42 (79)
NZ	41.0 (71)	42.0 (69)	41.4 (66)	—	44.7 (82)	—

Note: Each series for a column is based consistently on either pre- or post-tax income, but the basis of measurement may differ between the column years.

Sources: (1) Ahluwalia (1976); (2) Musgrave and Jarrett (1979); (3) Muller (1985); (4) World Bank, *World Development Report* (1979); (5) World Bank, *World Development Report* (1986; 1988); (6) OECD (1986).

TABLE 2.6. Estimates of percentage of work-force unionized 1950–1980

	1950	1955	1960	1965	1970	1975	1980
Aus.	62	63	63	63	62	59	58
Belg.	52	53	57	55	61	66	71
Den.	53	59	62	62	67	67	—
Fin.	33	31	32	36	57	75	—
Fr.	32	25	22	21	23	25	—
FRG	33	34	33	32	32	33	33
Gr.	—	—	—	—	—	—	—
Ice.	—	—	—	—	—	—	—
Ire.	—	—	—	—	—	—	—
It.	34	—	—	32	41	47	—
Lux.	—	—	—	—	—	—	—
Neth.	40	38	39	37	39	40	38
Norw.	52	54	63	64	64	61	65
Port.	—	—	—	—	—	—	—
Sp.	—	—	—	—	—	—	—
Sw.	67	68	68	71	73	82	85
Switz.	—	—	—	—	—	—	—
Turk.	—	—	—	—	—	—	—
UK	44	44	44	43	47	51	54
Can.	—	—	—	—	—	—	—
USA	28	31	29	27	26	23	23
Jap.	—	—	—	—	—	—	—
Austral.	56	58	55	54	51	54	53
NZ	—	—	—	—	—	—	—

Sources: Kjellberg (1983); Matheson (1979) for Finland.

TABLE 2.7. Estimates of work-force unionized (percentages)

	1946–60 (1)	1961–76 (2)	1978–80 (3)	1980 or nearest year (4)
Aus.	54	56	60	58
Belg.	42	52	72	75
Den.	48	50	71	79
Fin.	30	47	70	75
Fr.	28	19	20	22
FRG	36	34	40	33
Gr.	—	—	27	—
Ice.	—	—	—	80
Ire.	33	40	—	52
It.	27	18	50	37
Lux.	—	—	—	70
Neth.	31	33	39	38
Norw.	47	44	58	55
Port.	—	—	40	—
Sp.	—	—	37	—
Sw.	65	76	89	85
Switz.	25	22	38	35
Turk.	—	—	—	—
UK	43	44	56	54
Can.	25	27	—	39
USA	27	26	22	25
Jap.	26	28	27	32
Austral.	52	48	—	55
NZ	44	39	—	50

Sources: (1), (2): Korpi (1983); (3): Hartmann (1984); Smith (1984); Mielke (1983); (4): Therborn (1984).

TABLE 2.8. Migration

	Net migration rates		
	1970	1980	1985
Aus.	2.3	1.2	1.1
Belg.	0.4	−0.4	0.1
Den.	2.4	0.2	1.8
Fin.	−7.8	−0.6	0.6
Fr.	3.5	0.8	−0.1
FRG	9.3	5.1	1.5
Gr.	−4.4	5.0	1.1
Ice.	−7.3	−2.2	−2.1
Ire.	−1.0	−0.3	−7.3
It.	−0.9	0.1	1.5
Lux.	3.2	3.6	2.5
Neth.	2.6	3.7	1.4
Norw.	−0.3	1.0	1.4
Port.	−16.5	−12.9	2.3
Sp.	−0.6	0.0	0.4
Sw.	6.1	1.2	1.3
Switz.	−1.0	2.7	2.2
Turk.	—	0.1	−2.0
UK	−0.5	−0.9	1.3
Can.	3.3	4.5	1.1
USA	2.1	3.7	2.7
Jap.	−0.1	0.3	0.7
Austral.	9.8	7.1	5.2
NZ	3.9	−3.8	−4.2

Note: Net migration means immigration minus emigration.

Source: OECD (1988).

TABLE 2.9. Migration in Western Europe: minority labour force as percentage of total labour force

	1960 (1)	1970 (2)	1974 (3)	1977 (4)	1981 (5)	1982 (6)
Aus.	—	—	—	6.0	—	—
Belg.	4.8	7.1	—	8.4	8.7	8.2
Den.	—	—	—	—	—	—
Fin.	—	—	—	—	—	—
Fr.	—	6.3	8.2	7.3	6.4	6.6
FRG	1.7	6.5	10.9	9.5	9.5	9.2
Gr.	—	—	—	—	—	—
Ice.	—	—	—	—	—	—
Ire.	—	—	—	—	—	—
It.	—	—	—	—	—	—
Lux.	—	—	—	32.0	—	33.0
Neth.	1.1	2.8	3.2	3.0	4.9	3.7
Norw.	—	—	—	—	—	—
Port.	—	—	—	—	—	—
Sp.	—	—	—	—	—	—
Sw.	2.9	5.2	5.2	5.4	5.4	5.2
Switz.	21.8	25.2	25.4	16.4	22.9	19.7
UK	5.1	7.1	—	—	7.2	—

Sources: (1)–(3), (5): Castles (1984); (4): OECD, *Observer* (1979); (6): OECD, *Employment Outlook* (1985).

TABLE 2.10. Migration in Western Europe: minority population as percentage of total population

	1950 (1)	1960 (2)	1970 (3)	1975 (4)	1980 (5)	1982 (6)
Aus.	—	—	—	—	—	—
Belg.	4.3	4.9	7.2	8.5	9.2	8.9
Den.	—	—	—	—	—	—
Fin.	—	—	—	—	—	—
Fr.	4.5	5.4	6.5	7.9	7.7	6.8
FRG	1.1	1.2	4.9	6.6	7.2	7.6
Gr.	—	—	—	—	—	—
Ice.	—	—	—	—	—	—
Ire.	—	—	—	—	—	—
It.	—	—	—	—	—	—
Lux.	—	—	—	—	—	26.2
Neth.	1.1	1.0	1.8	2.6	3.4	3.8
Norw.	—	—	—	—	—	—
Port.	—	—	—	—	—	—
Sp.	—	—	—	—	—	—
Sw.	1.8	2.5	5.1	5.0	5.1	4.7
Switz.	6.1	10.8	15.8	16.0	14.2	14.5
UK	3.2	4.3	7.4	7.8	8.5	—

Sources: (1)–(5): Castles (1984); (6) OECD, Employment Outlook (1985).

TABLE 2.11. Migration in Western Europe: estimated population and labour force living abroad

	Estimated population living abroad			Estimated labour force living abroad	
	In Europe c.1970 (1)	In Europe c.1982 (2)	In world c.1982 (3)	In Europe c.1977 (4)	In Europe c.1982 (5)
Aus.	—	—	—	—	—
Belg.	—	—	—	—	—
Den.	—	—	—	—	—
Fin.	—	3.7	4.8	4.5	4.7
Fr.	—	—	—	—	—
FRG	—	—	—	—	—
Gr.	4.1	—	—	5.7	4.1
Ice.	—	—	—	—	—
Ire.	25.1	—	—	—	—
It.	3.4	4.1	9.4	4.0	4.6
Lux.	—	—	—	—	—
Neth.	—	—	—	—	—
Norw.	—	—	—	—	—
Port.	6.0	11.8	40.3	10.6	13.9
Sp.	2.9	2.3	5.6	3.2	3.1
Sw.	—	—	—	—	—
Switz.	—	—	—	—	—
UK	—	—	—	—	—

Sources: (1): Castles and Kosack (1973); (2), (3), (5): OECD, *Employment Outlook* (1985); (4): OECD, *Observer* (1979).

Section 3: Employment

Tables

3.1. Labour force as a percentage of total population
3.2. Female labour force as a percentage of total labour force
3.3. Employment in agriculture as a percentage of economically active population in civilian employment
3.4. Employment in industry as a percentage of economically active population in civilian employment
3.5. Employment in services as a percentage of economically active population in civilian employment
3.6. Armed forces as a percentage of total labour force
3.7. Public sector employment as a percentage of total civilian employment
3.8. Producers of government services as a percentage of total civilian employment
3.9. Government employment as a percentage of total civilian employment
3.10. Unemployment as a percentage of total labour force
3.11. Unemployment as a percentage of total labour force: five-year averages
3.12. Standardized unemployment rates as a percentage of total labour force, 1965–1985
3.13. Industrial disputes: workers involved per 1,000 in civilian labour force
3.14. Industrial disputes: working days lost per 1,000 in civilian labour force
3.15. Industrial disputes: number of persons involved in strikes per 1,000 of the non-agricultural labour force

Comparative Tables

TABLE 3.1. Labour force as a percentage of total population

	ILO figures				OECD figures			
	1950 (1)	1960 (2)	1970 (3)	1980 (4)	1960 (5)	1970 (6)	1980 (7)	1985 (8)
Aus.	47.9	48.1	41.6	45.1	47.0	42.0	43.9	44.4
Belg.	39.9	37.1	37.1	42.1	39.2	39.7	42.2	42.6
Den.	48.2	45.7	44.8	52.1	45.7	48.3	51.9	53.8
Fin.	49.3	45.7	46.1	45.9	48.6	47.7	51.7	52.9
Fr.	48.5	42.1	40.2	42.8	43.6	42.2	43.4	43.3
FRG	44.2	48.2	43.8	43.3	47.7	44.2	44.2	45.6
Gr.	37.2	42.3	36.9	36.4	43.3	37.2	35.8	39.1
Ice.	44.2	38.7	41.0	46.3	39.8	40.2	45.9	48.3
Ire.	45.1	38.4	37.6	36.9	39.4	37.9	36.7	36.7
It.	41.2	38.6	36.6	39.8	45.0	39.6	40.5	41.5
Lux.	46.4	41.2	36.6	42.0	41.9	41.2	43.6	44.6
Neth.	40.2	35.5	36.4	38.9	37.6	37.1	38.2	41.0
Norw.	42.4	39.1	37.7	49.9	40.6	40.1	47.6	49.7
Port.	39.4	37.9	38.4	43.5	37.6	39.9	46.8	46.6
Sp.	38.0	39.5	35.2	33.8	39.9	38.8	35.7	36.4
Sw.	44.2	41.8	41.6	48.2	49.0	48.7	52.0	53.0
Switz.	45.7	46.2	48.1	48.6	49.8	49.8	49.7	49.0
Turk.	60.6	47.2	42.8	42.5	48.6	41.8	39.3	37.7
UK	44.9	45.4	44.7	47.0	46.6	45.6	47.7	48.7
Can.	37.7	35.5	37.4	49.5	36.8	39.9	48.4	50.1
USA	40.0	38.8	42.0	46.3	39.9	41.9	47.7	49.0
Jap.	43.0	46.4	50.3	48.9	48.4	49.7	48.4	49.4
Austral.	44.7	40.1	41.1	45.9	40.0	43.4	45.7	46.6
NZ	38.6	37.1	38.8	42.0	36.9	38.7	41.4	42.7

Sources: (1)–(3): Mitchell (1981; 1982; 1983); Nordic Council (1984); (4): ILO, *Yearbook of Labour Statistics* (1981–5); Nordic Council (1984); (5)–(8): OECD (1985a; 1988). Historical statistics (1983; 1987).

Employment

TABLE 3.2. Female labour force as a percentage of total labour force

	1960	1970	1980	1985
Aus.	41.3	39.4	40.6	41.4
Belg.	30.2	32.0	37.7	39.8
Den.	30.9	38.6	44.1	45.6
Fin.	43.9	43.7	46.1	47.3
Fr.	32.5	35.0	39.7	41.8
FRG	37.3	35.9	37.8	38.7
Gr.	32.6	28.3	30.0	29.4
Ice.	32.6	32.5	30.0	29.4
Ire.	25.6	25.8	28.7	29.6
It.	30.7	28.8	33.3	34.8
Lux.	26.5	26.2	30.9	33.9
Neth.	21.5	23.5	30.3	34.0
Norw.	28.2	30.1	41.3	43.5
Port.	17.8	25.4	39.7	40.9
Sp.	21.7	23.4	28.5	30.0
Sw.	33.6	39.5	45.2	47.1
Switz.	34.2	34.0	36.1	37.0
Turk.	40.8	37.4	33.2	30.0
UK	32.7	35.5	39.2	40.4
Can.	26.6	33.3	39.9	42.4
USA	32.3	36.7	42.0	43.7
Jap.	40.7	39.3	38.7	39.7
Austral.	25.1	32.2	36.7	38.4
NZ	24.5	28.6	33.7	36.1

Source: OECD Historical statistics (1983; 1987).

TABLE 3.3. Employment in agriculture as a percentage of economically active population in civilian employment

	ILO figures				OECD figures			
	1950 (1)	1960 (2)	1970 (3)	1980 (4)	1960 (5)	1970 (6)	1980 (7)	1985 (8)
Aus.	32.5	22.8	13.8	8.5	22.6	14.5	9.0	8.2
Belg.	12.0	7.4	5.0	2.9	8.7	4.8	3.0	2.9
Den.	25.1	17.5	11.0	6.9	18.2	11.5	7.1	6.7
Fin.	45.9	35.5	20.3	12.3	35.2	22.6	13.5	11.5
Fr.	31.6	19.9	15.7	8.8	23.2	13.5	8.7	7.6
FRG	18.7	13.4	7.5	5.9	14.0	8.6	5.6	5.5
Gr.	48.2	55.3	40.5	28.1	57.1	40.8	30.3	28.9
Ice.	37.0	22.9	17.9	13.1	22.9	18.5	12.0	10.5
Ire.	39.6	35.2	25.4	15.6	37.3	27.1	18.3	16.0
It.	42.2	29.0	16.4	11.1	32.6	20.2	14.3	11.2
Lux.	26.0	14.9	7.5	5.0	16.6	9.3	5.4	4.2
Neth.	19.3	11.0	6.1	5.4	9.8	6.2	4.9	4.9
Norw.	25.9	19.5	11.6	8.1	21.6	13.9	8.5	7.2
Port.	48.5	43.0	30.3	19.2	43.9	30.0	27.3	23.2
Sp.	49.5	34.8	24.9	15.7	38.7	27.1	18.9	17.6
Sw.	19.1	13.8	6.8	5.6	15.7	8.1	5.6	4.8
Switz.	16.5	11.2	7.7	6.2	14.6	8.6	6.9	6.6
Turk.	85.3	75.0	67.7	57.6	75.9	67.6	60.7	57.1
UK	5.1	3.6	2.9	2.7	4.7	3.2	2.6	2.6
Can.	19.1	12.1	7.9	5.2	13.2	7.6	5.4	5.2
USA	11.9	6.5	4.3	3.0	8.5	4.5	3.6	3.1
Jap.	48.3	32.6	19.4	10.9	30.2	17.4	10.4	8.8
Austral.	13.4	10.9	7.4	6.0	11.0	8.0	6.5	6.2
NZ	17.4	14.4	12.3	10.8	14.6	12.1	10.9	11.1

Notes: Agriculture includes employment in agriculture, forestry, and fishing. Tables 3.3–3.5 do not always add up to 100% due to rounding.

Sources: See Table 3.1.

TABLE 3.4. Employment in industry as a percentage of economically active population in civilian employment

	ILO figures				OECD figures			
	1950 (1)	1960 (2)	1970 (3)	1980 (4)	1960 (5)	1970 (6)	1980 (7)	1985 (8)
Aus.	36.6	40.9	41.8	41.0	41.7	42.3	39.7	36.5
Belg.	48.8	47.0	43.1	34.2	45.0	42.0	34.1	29.7
Den.	33.3	36.5	37.2	31.2	36.9	37.8	30.4	28.1
Fin.	27.7	31.5	34.3	33.5	32.6	34.6	34.6	31.9
Fr.	32.9	38.4	40.4	35.9	38.4	39.2	35.9	32.0
FRG	44.9	47.2	47.9	44.6	47.0	48.5	44.1	41.0
Gr.	19.4	19.7	32.4	30.0	17.4	25.0	30.2	27.3
Ice.	30.6	33.8	38.1	37.9	34.7	36.9	38.2	36.8
Ire.	24.3	25.4	31.3	35.0	23.7	29.9	32.5	28.9
It.	32.1	40.4	42.2	39.5	33.9	39.5	37.9	33.6
Lux.	39.5	43.1	34.4	32.6	44.9	43.9	38.1	33.4
Neth.	36.9	43.4	36.7	30.6	40.5	38.9	31.4	26.5
Norw.	36.4	36.5	37.3	29.3	35.6	37.3	29.7	27.8
Port.	24.9	28.7	30.9	38.9	31.3	32.9	36.6	35.3
Sp.	25.5	35.1	37.4	35.9	30.3	35.5	36.1	31.8
Sw.	38.2	45.1	40.8	32.1	40.3	38.4	32.2	29.9
Switz.	46.7	50.5	48.2	39.0	46.5	45.9	39.7	37.7
Turk.	7.4	9.8	12.2	16.9	10.7	14.5	16.2	17.5
UK	49.1	47.4	37.9	38.1	47.7	44.7	37.7	32.3
Can.	35.5	33.2	30.8	28.1	32.7	30.9	28.5	25.5
USA	34.6	35.1	33.8	30.8	35.3	34.4	30.5	28.0
Jap.	22.6	29.7	34.4	34.2	28.5	35.7	35.3	34.9
Austral.	40.6	39.2	34.3	27.5	38.9	37.0	31.0	27.7
NZ	32.0	36.7	34.4	31.3	38.7	39.4	33.7	32.4

Note: Industry includes employment in mining and quarrying, manufacturing, electricity, gas and water, and construction. Tables 3.3–3.5 do not always add up to 100% due to rounding.

Sources: See Table 3.1.

TABLE 3.5. Employment in services as a percentage of economically active population in civilian employment

	ILO figures				OECD figures			
	1950 (1)	1960 (2)	1970 (3)	1980 (4)	1960 (5)	1970 (6)	1980 (7)	1985 (8)
Aus.	30.8	36.2	44.4	50.5	35.7	43.2	51.4	55.3
Belg.	39.2	45.5	51.9	62.9	46.4	53.2	62.9	67.4
Den.	41.6	45.9	51.7	61.9	44.8	50.7	62.4	65.2
Fin.	26.3	33.1	45.5	54.2	32.2	42.8	51.8	56.5
Fr.	35.4	41.6	43.9	55.3	38.5	47.2	55.4	60.4
FRG	36.4	39.3	44.6	49.5	39.1	42.9	50.3	53.5
Gr.	32.5	25.0	33.0	41.9	25.5	34.2	39.5	43.8
Ice.	32.4	43.3	44.0	49.0	42.4	44.6	49.8	52.7
Ire.	36.1	39.4	43.2	49.4	39.0	43.1	49.2	55.1
It.	25.7	30.6	36.6	49.4	33.5	40.3	47.8	55.2
Lux.	34.5	42.0	58.1	62.4	38.4	46.8	56.5	62.3
Neth.	43.8	48.0	57.3	64.0	49.7	54.9	63.6	68.6
Norw.	37.7	44.0	51.1	62.6	42.9	48.8	61.8	65.0
Port.	26.7	28.3	38.8	41.9	24.8	37.1	36.1	41.5
Sp.	25.0	30.1	37.7	48.4	31.0	37.4	45.1	50.6
Sw.	42.7	41.1	52.4	62.3	44.0	53.5	62.2	65.3
Switz.	37.0	38.3	43.9	54.8	38.9	45.5	53.4	55.7
Turk.	7.4	15.3	20.1	25.5	10.7	14.5	23.1	25.4
UK	45.9	48.9	59.2	59.2	47.6	52.0	59.7	65.1
Can.	45.4	54.7	61.3	66.7	54.1	61.4	66.0	69.3
USA	53.5	58.4	61.9	66.2	56.2	61.1	65.9	68.8
Jap.	29.1	37.7	46.2	54.9	41.3	46.9	54.2	56.4
Austral.	46.0	49.9	58.4	66.5	50.1	55.0	62.4	66.2
NZ	50.6	48.9	53.3	57.9	46.8	48.6	55.4	56.5

Notes: Services include employment in wholesale and retail trade, restaurants and hotels, transport, storage and communication, financing, insurance, real estate and business services, community and social and personal services, and related services. Tables 3.3–3.5 do not always add up to 100% due to rounding.

Sources: See Table 3.1.

TABLE 3.6. Armed forces as a percentage of total labour force

	1965	1970	1975	1980	1985
Aus.	—	—	—	—	—
Belg.	2.8	2.5	2.2	2.2	2.1
Den.	2.1	2.0	1.3	1.1	1.1
Fin.	1.7	1.9	1.7	1.7	1.5
Fr.	2.9	2.7	2.6	2.4	2.3
FRG	1.7	1.9	2.0	2.0	1.9
Gr.	—	—	—	—	—
Ice.	—	—	—	—	—
Ire.	0.7	0.7	1.0	1.2	1.3
It.	2.5	2.7	2.4	2.5	2.6
Lux.	0.8	0.7	0.6	0.6	0.6
Neth.	2.6	2.3	2.1	2.0	1.8
Norw.	3.2	3.1	—	—	—
Port.	4.5	4.4	3.2	2.0	1.6
Sp.	3.3	3.3	3.3	3.4	2.8
Sw.	—	—	—	—	—
Switz.	—	—	—	—	—
Turk.	3.4	3.4	3.0	2.8	2.7
UK	1.7	1.5	1.3	1.2	1.2
Can.	1.5	1.1	0.8	0.6	0.6
USA	2.5	2.5	1.8	1.5	1.5
Jap.	—	—	—	—	—
Austral.	1.2	1.5	1.1	1.1	1.0
NZ	1.1	1.1	0.9	0.8	0.9

Notes: The armed forces cover personnel from the metropolitan territory drawn from the total available labour force who were serving in the armed forces whether stationed in the metropolitan territory or elsewhere.

Source: OECD (1988).

TABLE 3.7. Public sector employment as a percentage of total civilian employment

	Total public sector		General government		Public enterprises	
	1970–4	1975–9	1970–4	1975–9	1970–4	1975–9
Aus.	27.9	30.2	15.0	17.8	12.9	12.4
Belg.	19.5	21.6	14.3	16.3	5.1	5.2
Den.	—	27.1	19.1	23.8	—	3.3
Fin.	—	—	—	—	—	—
Fr.	18.5	19.6	14.0	15.2	4.3	4.4
FRG	20.0	22.2	12.1	14.3	7.8	7.9
Gr.	—	—	—	—	—	—
Ice.	—	—	—	—	—	—
Ire.	18.0	19.7	11.9	14.0	6.3	5.7
It.	17.6	20.3	12.0	13.9	5.6	6.4
Lux.	13.5	14.3	9.6	10.6	3.9	3.7
Neth.	—	—	—	—	—	—
Norw.	21.7	23.9	17.6	20.0	4.1	4.2
Port.	—	—	—	—	—	—
Sp.	—	—	—	—	—	—
Sw.	30.0	34.5	22.8	27.7	7.1	8.0
Switz.	—	—	—	—	—	—
Turk.	—	—	—	—	—	—
UK	27.1	29.6	19.1	21.4	7.9	8.2
Can.	24.3	24.4	19.9	19.9	4.4	4.5
USA	19.3	18.8	17.7	17.2	1.6	1.6
Jap.	—	—	—	—	—	—
Austral.	23.2	25.5	—	—	—	—
NZ	23.1	24.3	17.8	18.7	5.3	5.6

Notes: Data are based on production sectors rather than institutional sectors; data for general government refer to producers of government services. Total public sector employment comprises general government employment and employment in public enterprises.

Sources: Pathirane and Blades (1982: 272-3).

TABLE 3.8. Producers of government services as a percentage of total civilian employment

	1970	1975	1980	1985
Aus.	—	—	—	—
Belg.	13.9	15.2	18.0	19.9
Den.	16.8	23.6	28.1	29.8
Fin.	11.8	14.7	18.2	20.4
Fr.	16.5	17.5	18.1	19.6 (1984)
FRG	11.2	13.9	14.9	16.1
Gr.	—	—	—	—
Ice.	—	13.9	15.7	16.4 (1984)
Ire.	—	—	—	—
It.	11.8	14.1	15.1	15.8
Lux.	9.4	9.7	10.8	—
Neth.	12.1	13.5	14.9	16.1
Norw.	16.4	19.3	21.7	23.8
Port.	7.3	8.1	8.8	—
Sp.	—	—	9.6	12.4
Sw.	20.5	25.5	30.7	33.2
Switz.	—	—	—	—
Turk.	—	—	—	—
UK	18.0	20.9	21.3	21.8
Can.	—	—	—	—
USA	18.1	18.0	16.7	15.4
Jap.	5.8	6.5	6.6	6.4
Austral.	—	8.7	4.5	4.8
NZ	—	—	—	—

Notes: Producers of government services mean all departments, establishments, and other bodies of central, state, and local governments which engage in different activities, i.e. 'producers of government services' is somewhat narrower than the field covered by general government.

Sources: 1970–1980: OECD, *National Accounts* (1983); 1985: OECD, *National Accounts* (1987).

TABLE 3.9. Government employment as a percentage of total civilian employment

	1960	1970	1975	1980	1985
Aus.	10.5	13.7	16.4	18.2	20.5
Belg.	12.2	13.9	15.7	18.6	19.9
Den.	—	16.8	23.6	28.3	29.8
Fin.	7.7	11.8	14.6	17.8	20.4
Fr.	13.1	13.4	14.3	15.6	17.8
FRG	8.0	11.2	13.9	14.9	16.0
Gr.	—	—	—	—	—
Ice.	—	—	13.9	15.7	16.4 (1984)
Ire.	—	11.2	13.3	14.4	16.1
It.	8.7	11.8	14.0	15.0	15.8
Lux.	—	9.4	9.7	10.8	11.2 (1984)
Neth.	11.7	12.1	13.5	14.9	16.1
Norw.	—	16.4	19.3	21.9	23.8
Port.	3.9	6.8	8.1	8.8	—
Sp.	—	7.1	10.0	11.9	14.3
Sw.	12.8	20.6	25.5	30.7	33.1
Switz.	6.4	7.9	9.5	10.7	11.2
Turk.	—	—	—	—	—
UK	14.8	18.0	20.7	21.1	21.8
Can.	—	19.5	20.3	18.8	20.0 (1984)
USA	15.7	18.1	18.0	16.7	15.8
Jap.	—	5.8	6.5	6.6	6.4
Austral.	22.3	22.9	25.5	25.4	26.4
NZ	17.9	18.2	18.9	19.0	18.1

Note: Government employment is a close approximation to the institutional sector 'general government'.

Sources: 1970–80: OECD, *Economic Outlook: Historical Statistics* (1983); 1985: OECD, *Economic Outlook: Historical Statistics* (1988).

Employment

TABLE 3.10. Unemployment as a percentage of total labour force

	1960	1970	1975	1980	1985
Aus.	2.4	1.4	1.7	1.6	4.1
Belg.	3.3	1.8	4.4	7.7	12.0
Den.[a]	1.9	0.7	4.9	6.5	7.3
Fin.	1.4	1.9	2.2	4.6	4.8
Fr.	1.2	2.4	4.1	6.3	10.2
FRG	1.0	0.6	4.0	3.3	8.3
Gr.[a]	6.1	4.2	2.3	2.8	7.8
Ice.[a]	—	1.2	1.1	1.0	0.9
Ire.[a]	5.6	5.8	7.3	7.3	17.4
It.	5.5	5.3	5.8	7.5	9.9
Lux.[a]	—	0.7	0.6	0.6	1.9
Neth.	0.7	1.0	5.2	6.0	12.8
Norw.	1.2	1.6	2.3	1.7	2.5
Port.[a]	1.9	2.5	4.4	7.7	9.0
Sp.	2.3	2.5	4.3	12.3	21.5
Sw.	1.7	1.5	1.6	2.0	2.8
Switz.	—	0.0	0.3	0.2	0.9
Turk.[a]	9.2	12.0	12.9	14.4	15.9
UK	1.3	2.2	3.2	5.6	11.5
Can.	6.4	5.6	6.9	7.4	10.4
USA	5.3	4.8	8.3	7.0	7.1
Jap.	1.7	1.1	1.9	2.0	2.6
Austral.	1.4	1.6	4.8	6.0	8.2
NZ[a]	0.1	0.2	0.2	2.2	4.1

[a] Denotes a different national definition of unemployment, whereas for the other countries OECD standards are adhered to.

Sources: 1960–70: OECD, *Economic Outlook: Historical Statistics* (1983); OECD (1985b); 1975–85: OECD, *Economic Outlook: Historical Statistics* (1987).

TABLE 3.11. Unemployment as a percentage of total labour force: five-year averages

	1950–4	1955–9	1960–4	1965–9	1970–4	1975–9	1980–4
Aus.	5.2	3.6	2.1	1.9	1.3	1.7	2.9
Belg.	4.9	3.2	2.2	2.2	2.1	6.2	11.1
Den.	4.4	4.4	1.8	1.7	1.9	6.6	9.5
Fin.	1.4	1.9	1.4	2.5	2.2	5.0	5.1
Fr.	1.3	1.1	1.2	1.6	2.1	4.9	8.0
FRG	6.7	3.1	0.7	0.9	1.1	3.8	6.2
Gr.	—	—	—	5.2	2.7	1.9	5.7
Ice.	—	—	—	1.5	1.1	1.0	1.2
Ire.	6.1	5.9	5.0	4.9	5.9	8.1	11.6
It.	7.4	6.9	3.1	3.6	3.2	6.8	8.6
Lux.	—	—	—	—	—	0.6	1.2
Neth.	2.4	1.5	0.9	1.4	2.1	5.3	9.9
Norw.	0.8	1.2	1.1	0.9	0.8	1.9	2.5
Port.	—	—	—	2.5	2.3	6.8	7.9
Sp.	1.2	0.8	1.2	1.6	2.5	5.9	15.6
Sw.	2.3	2.1	1.3	1.6	1.8	1.9	2.9
Switz.	—	—	—	—	—	0.4	0.6
Turk.	—	—	—	10.1	12.1	12.4	15.1
UK	1.2	1.2	1.5	1.7	2.5	4.6	9.5
Can.	3.2	5.0	5.9	4.2	5.9	7.5	9.8
USA	3.9	4.8	5.5	3.7	5.2	6.9	8.2
Jap.	2.0	2.2	1.4	1.2	1.3	2.0	2.4
Austral.	1.1	1.5	2.9	1.5	1.9	5.5	7.5
NZ	1.1	1.0	1.2	1.2	1.5	0.9	4.1

Sources: 1950–74: Madsen and Paldam (1978); 1975–84: OECD (1988).

TABLE 3.12. Standardized unemployment rates as a percentage of total labour force, 1965–1985

	1965	1966	1967	1968	1969	1970	1971	1972	1973	1974	1975	1976	1977	1978	1979	1980	1981	1982	1983	1984	1985
Aus.	1.9	1.8	1.9	2.0	2.0	1.4	1.3	1.2	1.1	1.4	1.7	1.8	1.6	2.1	2.1	1.9	2.5	3.5	4.1	3.8	3.6
Belg.	1.8	2.0	2.6	3.1	2.3	2.1	2.1	2.7	2.7	3.0	5.0	6.4	7.4	7.9	8.2	8.8	10.5	12.6	13.9	14.0	13.2
Fin.	1.4	1.5	2.9	3.8	2.8	1.9	2.2	2.5	2.3	1.7	2.2	3.8	5.8	7.2	5.9	4.6	5.1	5.8	6.1	6.1	6.2
Fr.	1.5	1.8	1.9	2.6	2.3	2.4	2.6	2.7	2.6	2.8	4.0	4.4	4.9	5.2	5.9	6.3	7.3	8.1	8.3	9.7	10.1
FRG	0.3	0.2	1.3	1.5	0.9	0.8	0.9	0.8	0.8	1.6	3.6	3.7	3.6	3.5	3.2	3.0	4.4	6.1	8.0	8.5	8.6
It.	5.3	5.7	5.3	5.6	5.6	5.3	5.3	6.3	6.2	5.3	5.8	6.6	7.0	7.1	7.6	7.5	8.3	9.0	9.8	10.2	10.5
Neth.	0.6	0.8	1.6	1.5	1.0	1.0	1.3	2.2	2.2	2.7	5.2	5.5	5.3	5.3	5.4	6.0	8.5	11.4	13.7	14.0	13.0
Norw.	1.8	1.6	1.5	2.1	2.0	1.6	1.5	1.7	1.5	1.5	2.3	1.8	1.5	1.8	2.0	1.7	2.0	2.6	3.3	3.0	2.5
Sp.	2.5	2.1	2.5	3.0	2.6	2.4	3.1	3.1	2.5	2.6	3.7	4.7	5.2	6.9	8.5	11.2	14.0	15.9	17.4	20.1	21.5
Sw.	1.2	1.6	2.1	2.2	1.9	1.5	2.5	2.7	2.5	2.0	1.6	1.6	1.8	2.2	2.1	2.0	2.5	3.1	3.5	3.1	2.8
Switz.	—	—	—	—	—	—	—	—	—	—	0.4	0.7	0.4	0.3	0.3	0.2	0.2	0.4	0.9	1.1	1.0
UK	2.3	2.3	3.4	3.4	3.1	3.0	3.7	4.0	3.0	2.9	4.3	5.7	6.1	6.0	5.1	6.6	9.9	11.4	12.6	13.0	13.1
Can.	3.6	3.3	3.8	4.4	4.4	5.6	6.1	6.2	5.5	5.3	6.9	7.1	8.0	8.3	7.4	7.4	7.5	10.9	11.8	11.2	10.4
USA	4.4	3.6	3.7	3.4	3.4	4.8	5.8	5.5	4.8	5.5	8.3	7.6	6.9	6.0	5.8	7.0	7.5	9.5	9.5	7.4	7.1
Jap.	1.2	1.3	1.3	1.2	1.1	1.1	1.2	1.4	1.3	1.4	1.9	2.0	2.0	2.2	2.1	2.0	2.2	2.4	2.6	2.7	2.6
Austral.	1.5	1.7	1.9	1.8	1.8	1.6	1.9	2.6	2.3	2.6	4.8	4.7	5.6	6.2	6.2	6.0	5.7	7.1	9.9	8.9	8.2

Notes: By standardized unemployment rates is meant that efforts have been made to make rates for different countries comparable.

Sources: 1965–9: OECD, *Economic Outlook* (1981); 1970–85: OECD, *Economic Outlook* (1986).

TABLE 3.13. Industrial disputes: workers involved per 1,000 in civilian labour force

	1960–4	1965–9	1970–4	1975–9	1980–4
Aus.	20	18	6	0	2
Belg.	6	8	20	22	7[a]
Den.	19	9	46	31	19
Fin.	19	19	173	156	162
Fr.	108	188	102	67	12
FRG	3	3	9	41	7
Gr.	25	68[b]	—	159	196
Ice.	108	185	114	274	108
Ire.	15	39	31	33	24
It.	141	185	255	605	418
Lux.	—	—	—	—	—
Neth.	6	2	7	4	6
Norw.	5	0	4	4	8
Port.	—	—	—	70[c]	68
Sp.	6[d]	10	27	241	137
Sw.	1	2	6	4	41
Switz.	0	0	0	0	0
Turk.	0[d]	1	1	2	1
UK	61	48	62	65	49
Can.	12	35	48	60	29
USA	19	30	31	12	7
Jap.	28	25	42	22	5
Austral.	102	134	230	224	121
NZ	27	30	80	119	109

[a] 1980.
[b] 1965–7.
[c] 1977 and 1979.
[d] 1963 and 1964.

Sources: 1960–74: Mitchell (1981; 1982; 1983); OECD (1985); 1975–84: ILO, *Yearbook of Labour Statistics* (1984; 1985); OECD (1988).

TABLE 3.14. Industrial disputes: working days lost per 1,000 in civilian labour force

	1960–4	1965–9	1970–4	1975–9	1980–4
Aus.	55	17	14	1	1
Belg.	79	72	269	191	53[a]
Den.	230	31	356	67	86
Fin.	150	84	587	362	375
Fr.	164	297	162	165	72
FRG	18	6	47	41	43
Gr.	61	127[b]	—	297	463
Ice.	1 394	1 179	1 022	1 286	938
Ire.	242	517	405	621	310
It.	400	755	972	994	611
Lux.	—	—	—	—	—
Neth.	27	5	45	23	18
Norw.	103	7	47	26	53
Port.	—	—	—	99[c]	124
Sp.	11[d]	23	85	947	379
Sw.	5	25	54	26	218
Switz.	5	0	1	2	1
Turk.	9[d]	20	37	130	154
UK	129	157	566	451	397
Can.	180	628	723	782	535
USA	263	471	499	216	132
Jap.	97	63	113	55	10
Austral.	154	211	570	474	366
NZ	64	102	186	303	286

[a] 1980.
[b] 1965–7.
[c] 1977 and 1979.
[d] 1963 and 1964.

Sources: 1960–74: Mitchell (1981; 1982; 1983); OECD (1985); 1975–84: ILO, *Yearbook of Labour Statistics* (1984; 1985); OECD (1985).

TABLE 3.15. Industrial disputes: number of persons involved in strikes per 1,000 of the non-agricultural labour force

	1919–38	1946–52	1960–7	1968–73	1974–7	1978–82
Aus.	7	5	17	3	1	1
Belg.	31	21	8	17	21	15
Den.	5	9	8	14	31	29
Fin.	6	50	18	47/230[a]	219	97
Fr.	22	94	127	135	110	27
FRG	26	5	2	4	4	5
Gr.	—	—	—	—	—	—
Ice.	—	—	—	—	—	—
Ire.	12	25	27	44	43	36
It.	6	199	135	248	371	272
Lux.	—	—	—	—	—	—
Neth.	8	3	3	5	2	2
Norw.	21	3	1	1	5	4
Port.	—	—	—	—	—	—
Sp.	—	—	—	—	—	—
Sw.	19	9	1	2	5	11
Switz.	3	0.2	0.2	0.1	0.3	0.2
Turk.	—	—	—	—	—	—
UK	24	39	39	69	41	70
Can.	12	25	19	38	56	33
USA	19	28	23	33	26	14
Jap.	3	34	35	40	40	10
Austral.	43	149	108	216	251	210
NZ	16	49	26	69	102	106

[a] Denotes a break in the statistical series for Finland.

Sources: Therborn (1984).

Section 4: Economy

TABLES

4.1. GDP per capita in US $ (constant prices; US $ 1980 at 1980 exchange rate)
4.2. Real GDP per capita in international prices (constant prices; US $ 1980 at 1980 exchange rate)
4.3. Origin of GDP by sector: agriculture (percentage)
4.4. Origin of GDP by sector: industry (percentage)
4.5. Origin of GDP by sector: services (percentage)
4.6. Inflation: average growth rates 1950–1980
4.7. Inflation: average growth rates 1960–1985
4.8. External dependency: exports as a percentage of GDP
4.9. External dependency: imports as a percentage of GDP
4.10. External dependency: imports and exports as a percentage of GDP
4.11. External dependency: trade balance as a percentage of GDP
4.12. Growth of real GDP 1960–1985 (year to year averages)
4.13. Growth of real GDP per capita 1960–1985 (year to year averages)
4.14. Growth of real GDP 1950–1981 (year to year averages)
4.15. Growth of real GDP per capita 1950–1981 (year to year averages)
4.16. Growth of GNP per capita 1960–1981 (year to year averages)

TABLE 4.1. GDP per capita in US $ (constant prices; US $ 1980 at 1980 exchange rate)

	1960	1965	1970	1975	1980	1985
Aus.	4 798	5 694	7 176	8 573	10 184	10 936
Belg.	5 796	7 156	8 888	10 378	11 985	12 431
Den.	7 492	9 324	10 784	11 580	12 941	14 635
Fin.	5 149	6 339	7 916	9 455	10 815	12 220
Fr.	5 920	7 339	9 156	10 731	12 335	12 972
FRG	7 215	8 669	10 276	11 165	13 216	14 198
Gr.	1 460	2 093	2 882	3 584	4 164	4 313
Ice.	6 472	8 137	8 536	10 849	14 164	14 675
Ire.	2 861	3 411	4 189	4 845	5 656	5 954
It.	3 831	4 766	6 244	6 809	8 081	8 647
Lux.	7 838	8 873	10 230	11 259	12 489	14 325
Neth.	6 756	7 986	9 751	10 893	11 970	12 249
Norw.	6 723	8 116	9 364	11 374	14 121	16 401
Port.	999	1 349	1 849	2 212	2 701	2 734
Sp.	2 370	3 391	4 343	5 420	5 668	5 907
Sw.	8 689	10 816	12 718	14 187	14 938	16 250
Switz.	10 560	12 294	14 329	14 600	15 920	16 642
Turk.	727	811	984	1 230	1 272	1 442
UK	6 398	7 212	7 964	8 763	9 495	10 337
Can.	5 687	6 837	7 897	9 553	10 934	11 902
USA	7 773	9 141	9 970	10 586	11 804	13 048
Jap.	2 659	4 084	6 491	7 456	9 069	10 606
Austral.	6 064	6 963	8 325	9 350	10 064	11 001
NZ	5 573	6 247	6 601	7 387	7 123	7 797

Sources: OECD, *National Accounts 1960–1988* (1988): vol. i, part vi, table 20.

Economy

TABLE 4.2. Real GDP per capita in international prices (constant prices; US $ 1980 at 1980 exchange rate)

	1950	1960	1970	1980	1985
Aus.	2 318	3 908	5 843	8 230	8 929
Belg.	3 462	4 379	6 750	9 228	9 717
Den.	4 241	5 490	7 776	9 598	10 884
Fin.	2 758	4 073	6 186	8 393	9 232
Fr.	3 125	4 473	7 078	9 688	9 918
FRG	2 713	5 217	7 443	9 795	10 708
Gr.	986	1 474	2 952	4 383	4 464
Ice.	3 592	4 644	6 157	9 285	9 037
Ire.	2 047	2 545	3 628	4 929	5 205
It.	1 929	3 233	5 028	7 164	7 425
Lux.	5 286	6 112	7 857	10 173	10 540
Neth.	3 404	4 690	6 915	9 036	9 092
Norw.	3 802	5 001	7 104	11 094	12 623
Port.	937	1 429	2 575	3 733	3 729
Sp.	1 640	2 425	4 379	6 131	6 437
Sw.	3 980	5 149	7 401	8 863	9 904
Switz.	4 886	6 834	9 164	10 013	10 640
Turk.	822	1 255	1 702	2 319	2 533
UK	3 993	4 970	6 319	7 975	8 665
Can.	5 337	6 069	8 495	11 332	12 196
USA	6 401	7 380	9 459	11 404	12 532
Jap.	1 129	2 239	5 496	8 117	9 447
Austral.	4 331	5 182	7 344	8 349	8 850
NZ	4 531	5 571	6 595	7 363	8 000

Note: Real GDP per capita is adjusted for differences in the purchasing power of currencies, i.e. efforts have been made to make better international comparisons.

Sources: Summers and Heston (1988): tables 53, 70–1, 73–81, 83–90, 92, 104, 117, 120.

TABLE 4.3. Origin of GDP by sector: agriculture (percentage)

	1950	1960	1970	1980	1985
Aus.	18	11.0	6.9	4.4	3.4
Belg.	9	6.5	3.6	2.1	2.4
Den.	21	11.2	5.6	4.5	4.8
Fin.	26	16.7	11.4	8.2	7.1
Fr.	15	10.6	6.5	4.2	3.8
FRG	10	5.7	3.4	2.2	1.7
Gr.	31	22.9	18.3	17.4	15.3
Ice.	—	—	—	10.1	7.8
Ire.	29	—	14.4	13.3	—
It.	22	12.2	8.1	6.4	4.2
Lux.	—	6.6	3.8	2.5	2.7
Neth.	14	—	5.8	3.5	4.1
Norw.	14	9.0	6.4	4.5	3.1
Port.	33	24.8	18.0	12.7	9.4
Sp.	—	—	10.5	7.1	6.2
Sw.	12	—	4.1	3.2	2.9
Switz.	—	—	—	—	—
Turk.	49	40.8	26.7	21.4	17.4
UK	6	3.4	2.5	1.9	1.6
Can.	13	5.7	3.7	3.9	2.6
USA	7	4.0	2.7	2.8	2.1
Jap.	26	12.6	6.1	3.8	3.1
Austral.	24	12.0	6.1	6.8	4.0
NZ	—	—	12.6	11.3	11.1

Note: Country totals in Tables 4.3–4.5 do not always add up to 100% because of rounding.

Sources: 1950: Mitchell (1981; 1982; 1983); 1960–80: World Bank (1984): Series 1; 1985: World Bank (1988).

Economy

TABLE 4.4. Origin of GDP by sector: industry (percentage)

	1950	1960	1970	1980	1985
Aus.	48	46.6	45.4	39.9	38.2
Belg.	44	40.9	42.3	36.7	33.1
Den.	36	31.2	29.6	23.0	23.6
Fin.	40	34.5	35.5	35.7	32.5
Fr.	48	39.1	38.8	35.8	33.7
FRG	49	53.3	53.1	47.9	37.4
Gr.	22	25.8	31.4	31.3	26.2
Ice.	—	—	—	33.8	31.8
Ire.	25	—	31.4	34.2	—
It.	40	41.3	42.9	42.7	33.1
Lux.	—	51.4	54.1	39.2	39.4
Neth.	40	—	37.4	32.6	33.7
Norw.	37	32.5	33.8	40.4	41.1
Port.	35	36.4	41.6	45.9	39.0
Sp.	—	—	37.1	37.9	37.3
Sw.	44	—	35.9	31.5	30.6
Switz.	—	—	—	—	—
Turk.	15	20.5	24.5	28.6	32.7
UK	48	42.8	38.1	34.6	37.0
Can.	38	34.4	31.7	33.7	31.9
USA	39	38.3	34.9	34.1	31.1
Jap.	32	44.6	46.6	42.9	40.9
Austral.	35	39.9	40.5	36.4	33.4
NZ	—	—	31.5	31.3	33.0

Note: Country totals in Tables 4.3–4.5 do not always add up to 100% because of rounding.

Sources: 1950: Mitchell (1981; 1982; 1983); 1960–80: World Bank (1984): Series 1; 1985: World Bank (1988).

TABLE 4.5. Origin of GDP by sector: services (percentage)

	1950	1960	1970	1980	1985
Aus.	34	42.4	47.7	55.7	58.4
Belg.	47	52.6	54.1	61.2	64.5
Den.	43	57.6	64.8	72.5	71.6
Fin.	34	48.8	53.1	56.1	60.4
Fr.	37	50.3	54.7	60.0	62.5
FRG	41	41.0	43.5	49.9	60.9
Gr.	47	51.3	50.3	51.3	58.5
Ice.	—	—	—	56.1	60.4
Ire.	46	—	54.2	52.5	—
It.	38	46.5	49.0	50.9	62.7
Lux.	—	42.0	42.1	58.3	57.9
Neth.	46	—	56.8	63.9	62.2
Norw.	49	58.5	59.8	55.1	55.8
Port.	32	38.8	40.4	41.4	51.6
Sp.	—	—	52.4	55.0	56.5
Sw.	44	—	60.0	65.3	66.5
Switz.	—	—	—	—	—
Turk.	36	38.7	48.8	50.0	42.6
UK	46	53.8	59.4	63.5	61.4
Can.	49	59.9	64.6	62.4	65.5
USA	54	57.7	62.4	63.1	66.8
Jap.	42	42.8	47.3	53.3	56.0
Austral.	41	48.1	53.4	56.8	62.6
NZ	—	—	55.9	57.4	55.9

Note: Country totals in Tables 4.3–4.5 do not always add up to 100% because of rounding.

Sources: 1950: Mitchell (1981; 1982; 1983); 1960–80: World Bank (1984): Series 1; 1985: World Bank (1988).

TABLE 4.6. Inflation: average growth rates 1950–1980

	Consumer price index			Implicit GDP deflator		
	1950–60	1960–70	1970–80	1950–60	1960–70	1970–81
Aus.	5.3	3.5	6.5	5.8	3.7	6.1
Belg.	1.9	3.2	7.8	—	3.6	7.3
Den.	3.4	6.0	10.3	3.4	6.4	10.0
Fin.	5.6	3.5	12.0	6.2	6.0	12.0
Fr.	5.6	3.8	10.1	6.6	4.3	9.9
FRG	1.9	2.6	5.0	3.0	3.2	5.0
Gr.	5.8	2.2	15.4	6.2	3.2	14.8
Ice.	3.7	11.7	36.2	11.2	12.2	36.8
Ire.	3.8	4.6	14.5	3.8	5.2	14.2
It.	3.0	4.0	14.9	2.2	4.4	15.7
Lux.	1.6	2.7	7.0	—	3.7	6.8
Neth.	3.0	4.4	7.4	3.5	5.4	7.6
Norw.	4.4	4.2	8.8	4.3	4.4	8.8
Port.	1.0	9.5	19.7	1.9	3.0	17.0
Sp.	5.2	6.8	15.9	—	6.8	16.0
Sw.	3.2	4.1	9.5	4.9	4.3	10.0
Switz.	1.4	3.5	4.8	2.1	4.4	4.8
Turk.	9.5	6.1	32.9	10.1	5.6	32.7
UK	4.0	3.8	14.3	2.5	4.1	14.4
Can.	2.2	2.8	8.7	2.7	3.1	9.3
USA	2.1	2.6	8.1	2.5	2.9	7.2
Jap.	4.1	5.6	9.3	—	5.1	7.4
Austral.	6.0	2.5	11.1	3.9	3.1	11.5
NZ	4.3	3.7	13.0	4.2	3.6	12.9

Notes: The consumer price index measures changes in the cost of living; the implicit GDP deflator provides a comprehensive measure of the aggregate price movements of goods and services making up the GDP.

Source: World Bank (1984): Series 1.

TABLE 4.7. Inflation: average growth rates 1960–1985

	GDP implicit price index: year to year changes				
	1960–8	1968–73	1973–9	1979–85	1960–85
Aus.	3.9	5.8	6.0	4.9	5.0
Belg.	3.2	5.5	8.1	5.4	5.4
Den.	6.1	8.6	10.2	7.9	8.0
Fin.	6.3	7.6	12.5	8.9	8.6
Fr.	4.0	6.4	10.5	9.9	7.4
FRG	3.1	6.3	4.8	3.4	4.2
Gr.	3.0	6.8	15.5	19.9	10.6
Ice.	11.4	19.9	38.7	47.9	27.6
Ire.	4.5	11.4	14.4	11.6	9.9
It.	4.3	7.2	17.1	15.1	10.4
Lux.	2.6	7.3	6.4	7.3	5.5
Neth.	5.0	7.8	7.4	3.9	5.8
Norw.	4.0	7.5	8.2	9.5	7.0
Port.	2.6	6.3	19.7	21.5	11.7
Sp.	6.5	7.9	18.3	11.8	10.8
Sw.	4.2	5.9	10.6	9.0	7.2
Switz.	4.5	6.8	3.7	4.3	4.7
Turk.	5.4	14.9	30.8	46.9	22.3
UK	3.7	7.5	16.0	8.9	8.5
Can.	3.0	5.3	9.2	7.0	5.9
USA	2.5	5.3	8.0	5.6	5.1
Jap.	5.3	7.3	8.1	2.1	5.5
Austral.	2.6	6.8	12.2	9.0	7.2
NZ	3.0	8.8	13.7	11.6	8.7

Notes: The implicit GDP price index provides a comprehensive measure of the aggregate price movements of goods and services making up the GDP.

Source: OECD, *Economic Outlook: Historical Statistics* (1987).

TABLE 4.8. *External dependency: exports as a percentage of GDP*

	1950	1955	1960	1965	1970	1975	1980	1985
Aus.	15.2	21.1	24.0	25.8	32.4	33.7	38.7	40.3
Belg.	27.8	31.9	32.8	36.3	43.9	46.1	59.9	73.9
Den.	27.1	32.9	32.8	29.8	29.6	30.1	32.7	36.9
Fin.	20.0	21.7	23.5	21.3	26.2	24.3	34.2	29.8
Fr.	15.6	15.0	15.0	13.7	16.3	19.5	22.3	25.1
FRG	11.5	20.0	20.5	19.5	21.2	25.0	27.2	32.5
Gr.	5.3	11.5	9.3	9.3	10.0	16.3	19.7	21.2
Ice.	35.1	30.4	44.5	38.6	47.7	36.3	37.1	45.1
Ire.	28.6	28.2	31.5	31.8	37.0	43.4	53.2	66.7
It.	11.7	11.0	14.7	16.9	17.8	22.8	25.1	23.7
Lux.	94.4	80.7	88.9	79.8	89.4	91.0	82.9	84.9
Neth.	41.2	47.9	50.2	45.0	47.2	52.3	52.7	64.2
Norw.	46.7	40.7	41.1	40.0	41.8	41.8	47.6	47.2
Port.	22.2	19.6	16.9	25.9	23.5	19.7	28.1	37.3
Sp.	—	—	11.2	10.6	13.5	13.3	15.5	25.9
Sw.	25.7	26.9	25.0	23.2	24.3	28.2	30.0	35.2
Switz.	25.6	28.2	30.0	30.6	32.8	31.4	36.7	39.1
Turk.	—	—	6.2	7.3	5.9	—	—	—
UK	23.2	21.9	20.2	18.5	23.6	26.6	28.3	29.4
Can.	22.2	20.3	18.5	20.5	23.4	23.3	29.4	28.8
USA	4.3	4.4	4.8	4.9	5.7	8.5	10.2	7.1
Jap.	11.8	10.7	11.1	10.8	10.8	12.8	13.9	14.7
Austral.	—	—	14.9	17.8	14.9	15.1	16.9	16.0
NZ	—	—	—	—	26.7	23.4	29.0	31.2

Note: Figures in the 1985 column refer to 1984 for Ireland and Spain and to 1982 for Luxembourg.

Sources: 1950–65: OECD, *National Accounts 1950–1968* (1968); OECD, *National Accounts 1960–1977* (1979); 1970–80: OECD, *National Accounts 1964–1981* (1981); 1985: OECD, *National Accounts 1973–1985* (1987).

TABLE 4.9. External dependency: imports as a percentage of GDP

	1950	1955	1960	1965	1970	1975	1980	1985
Aus.	19.5	22.5	25.4	26.8	31.4	33.0	40.6	40.1
Belg.	28.4	31.1	33.9	36.2	33.2	37.1	50.8	72.2
Den.	30.5	31.9	33.8	31.1	32.4	31.0	33.8	36.7
Fin.	18.8	20.1	24.2	23.2	27.4	30.5	34.9	28.7
Fr.	14.4	13.0	13.2	12.9	15.8	18.8	24.1	25.0
FRG	12.8	17.5	18.1	19.4	19.1	22.4	27.7	29.0
Gr.	21.1	17.8	23.9	22.0	18.4	26.9	26.4	32.8
Ice.	40.3	33.0	47.9	36.6	45.1	44.7	36.4	44.4
Ire.	43.1	40.0	36.7	40.8	45.0	48.7	66.7	64.2
It.	13.2	12.1	14.9	14.2	17.2	22.7	28.0	24.2
Lux.	74.2	74.2	76.0	81.2	77.0	88.1	86.1	88.4
Neth.	48.4	47.0	48.1	45.6	49.0	48.8	53.2	59.3
Norw.	44.9	43.6	42.8	40.2	43.1	48.5	41.4	39.3
Port.	24.7	24.0	23.2	30.9	30.4	32.2	43.4	41.4
Sp.	—	—	8.2	14.0	14.4	17.2	18.1	23.4
Sw.	25.1	27.8	25.7	24.4	24.9	28.4	31.9	32.8
Switz.	25.9	26.8	29.5	30.7	34.5	28.6	40.3	38.6
Turk.	—	—	8.5	9.6	8.6	—	—	—
UK	23.8	23.4	21.8	19.4	22.5	28.2	25.9	28.2
Can.	21.7	21.3	20.0	20.9	20.6	24.8	27.5	26.0
USA	4.1	4.5	4.4	4.5	5.5	7.8	11.0	10.1
Jap.	11.4	10.1	10.6	9.4	9.5	12.8	14.9	11.2
Austral.	—	—	17.8	17.7	15.1	14.9	19.0	19.0
NZ	—	—	—	—	25.9	30.4	30.2	34.3

Note: Figures in the 1985 column refer to 1984 for Ireland and Spain and to 1982 for Luxembourg.

Sources: 1950–65: OECD, *National Accounts 1950–1968* (1968); OECD, *National Accounts 1960–1977* (1979); 1970–80: OECD, *National Accounts 1964–1981* (1981); 1985: OECD, *National Accounts 1973–1985* (1987).

TABLE 4.10. External dependency: imports and exports as a percentage of GDP

	1950	1955	1960	1965	1970	1975	1980	1985
Aus.	34.7	43.6	49.4	52.6	63.9	66.7	79.3	80.3
Belg.	56.1	63.0	66.7	72.5	77.0	83.2	110.7	146.1
Den.	57.6	64.9	66.5	60.9	62.0	61.1	66.5	73.6
Fin.	38.8	41.8	47.7	44.5	53.6	54.7	69.1	58.4
Fr.	30.0	27.9	28.2	26.6	32.2	38.4	46.4	50.1
FRG	24.3	37.4	38.6	39.0	40.3	47.4	54.8	61.5
Gr.	26.4	29.3	33.2	31.3	28.4	43.2	46.1	54.0
Ice.	75.4	63.4	92.4	75.2	92.7	81.0	73.5	89.5
Ire.	71.7	68.2	68.2	72.6	81.9	92.1	119.9	130.9
It.	24.8	23.0	29.6	31.1	35.0	45.5	53.0	47.9
Lux.	168.6	154.9	164.8	161.1	166.4	179.1	168.9	173.3
Neth.	89.6	94.9	98.3	90.6	96.2	101.0	105.9	123.4
Norw.	91.6	84.3	83.9	80.2	84.9	90.3	89.0	86.5
Port.	46.9	43.6	40.1	56.7	53.9	51.9	71.5	78.7
Sp.	—	—	19.3	24.6	27.9	0.5	33.6	49.3
Sw.	50.8	54.7	50.6	47.6	49.2	6.7	61.9	68.0
Switz.	51.4	55.0	59.5	61.3	67.2	60.0	77.0	77.8
Turk.	—	—	14.7	16.9	14.5	—	—	—
UK	47.0	45.3	42.0	37.9	46.1	54.8	54.2	57.6
Can.	43.9	41.6	38.5	41.5	44.0	48.1	56.9	54.8
USA	8.4	8.8	9.2	9.4	11.2	16.3	21.2	17.2
Jap.	23.2	20.9	21.6	20.2	20.4	35.6	28.8	25.9
Austral.	—	—	32.7	33.0	30.0	30.0	35.8	34.9
NZ	—	—	—	—	52.6	53.8	59.2	65.5

Note: Figures in the 1985 column refer to 1984 for Ireland and Spain and to 1982 for Luxembourg.

Sources: 1950–65: OECD, *National Accounts 1950–1968* (1968); OECD, *National Accounts 1960–1977* (1979); 1970–80: OECD, *National Accounts 1964–1981* (1981); 1985: OECD, *National Accounts 1973–1985* (1987).

TABLE 4.11. External dependency: trade balance as a percentage of GDP

	1950	1955	1960	1965	1970	1975	1980	1985
Aus.	-4.2	-1.4	-1.4	-1.0	1.0	0.6	-1.9	0.2
Belg.	-0.6	0.8	-1.1	0.2	10.7	8.9	9.1	1.8
Den.	-3.4	1.0	-1.0	-1.3	-2.7	-0.9	-1.1	0.2
Fin.	1.2	1.7	-0.8	-1.9	-1.3	-6.2	-0.8	1.1
Fr.	1.2	2.0	1.8	0.9	0.5	0.7	-1.7	0.2
FRG	-1.4	2.5	2.4	0.1	2.0	2.7	-0.5	3.6
Gr.	-15.8	-6.2	-14.6	-12.7	-8.4	-10.5	-6.8	-11.6
Ice.	-5.2	-2.6	-3.4	2.0	2.6	-8.4	0.6	0.7
Ire.	-14.5	-11.8	-5.3	-9.0	-8.0	-5.2	-13.5	2.4
It.	-1.5	-1.1	-0.3	2.6	0.5	0.0	-2.9	-0.5
Lux.	20.2	6.6	12.9	-1.4	12.4	2.9	-3.2	-3.5
Neth.	-7.3	0.8	2.1	-0.6	-1.8	3.5	-0.4	4.9
Norw.	1.8	-2.9	-1.7	-0.2	-1.3	-6.7	6.2	7.9
Port.	-2.4	-4.5	-6.3	-5.0	-6.8	-12.5	-15.4	-4.1
Sp.	—	—	3.0	-3.4	-0.9	-3.9	-2.6	2.5
Sw.	0.6	-0.9	-0.7	-1.2	-0.6	-0.2	-1.9	2.4
Switz.	-0.3	1.4	0.5	-0.1	-1.7	2.9	-3.5	0.5
Turk.	—	—	-2.4	-2.3	-2.7	—	—	—
UK	-0.6	-1.6	-1.6	-0.9	1.0	-1.6	2.4	1.2
Can.	0.5	-1.0	-1.5	-0.4	2.7	-1.5	1.9	2.8
USA	0.2	-0.1	0.3	0.4	0.2	0.8	-0.8	-3.0
Jap.	0.4	0.6	0.5	1.5	1.3	0.0	-0.9	3.4
Austral.	—	—	-2.9	-2.3	-0.2	0.2	-2.1	-3.0
NZ	—	—	—	—	0.8	-6.9	-1.2	-3.2

Note: Figures in the 1985 column refer to 1984 for Ireland and Spain and to 1982 for Luxembourg. By trade balance is meant exports minus imports.

Sources: 1950–65: OECD, *National Accounts 1950–1968* (1968); OECD, *National Accounts 1960–1977* (1979); 1970–80: OECD, *National Accounts 1964–1981* (1981); 1985: OECD, *National Accounts 1973–1985* (1987).

Economy

TABLE 4.12. Growth of real GDP 1960–1985 (year to year averages)

	1960–8 (1)	1968–73 (2)	1973–9 (3)	1979–85 (4)	1960–81 (5)	1960–85 (6)
Aus.	4.2	5.9	2.9	1.8	4.0	3.7
Belg.	4.5	5.6	2.2	1.2	3.7	3.4
Den.	4.6	4.0	1.9	1.9	3.2	3.2
Fin.	3.9	6.7	2.4	3.2	4.1	3.9
Fr.	5.4	5.9	3.1	1.1	4.4	3.9
FRG	4.1	4.9	2.3	1.3	3.6	3.1
Gr.	7.3	8.2	3.7	1.1	5.8	5.1
Ice.	4.1	6.9	5.6	1.5	4.7	4.4
Ire.	4.2	4.8	4.6	2.0	4.0	3.9
It.	5.7	4.6	2.6	1.4	4.2	3.7
Lux.	3.0	5.9	1.5	2.2	3.1	3.0
Neth.	4.8	4.9	2.7	0.7	3.8	3.3
Norw.	4.4	4.1	4.9	3.3	4.3	4.2
Port.	6.6	7.4	3.1	1.6	5.4	4.7
Sp.	7.5	6.8	2.5	1.4	5.3	4.7
Sw.	4.4	3.7	1.8	1.8	3.1	3.0
Switz.	4.4	4.5	−0.4	1.9	2.9	2.6
Turk.	5.8	5.5	6.6	4.0	5.3	5.5
UK	3.1	3.2	1.5	1.2	2.1	2.3
Can.	5.4	5.4	4.2	2.4	4.0	4.4
USA	4.4	3.2	2.4	2.5	3.3	3.2
Jap.	10.4	8.4	3.6	4.0	7.5	6.8
Austral.	5.0	5.5	2.7	2.9	4.2	4.0
NZ	3.1	5.1	0.2	2.6	2.9	2.7

Sources: (1)–(4), (6): OECD, *Economic Outlook: Historical Statistics* (1987): table 3.1; (5): OECD, *Economic Outlook: Historical Statistics* (1983): table 3.1.

TABLE 4.13. Growth of real GDP per capita 1960–1985
(year to year averages)

	1960–8 (1)	1968–73 (2)	1973–9 (3)	1979–85 (4)	1960–81 (5)	1960–85 (6)
Aus.	3.6	5.4	3.0	1.8	3.7	3.4
Belg.	3.9	5.3	2.0	1.2	3.4	3.1
Den.	3.8	3.3	1.6	1.9	2.7	2.7
Fin.	3.3	6.5	2.0	2.7	3.7	3.5
Fr.	4.2	5.0	2.6	0.6	4.4	3.1
FRG	3.2	4.0	2.5	1.4	3.6	2.7
Gr.	6.7	7.8	2.6	0.4	5.0	4.3
Ice.	2.4	5.8	4.5	0.3	3.3	3.1
Ire.	3.8	3.7	3.0	1.1	3.1	2.9
It.	5.0	3.9	2.1	1.2	3.5	3.2
Lux.	2.1	5.0	0.9	2.1	2.3	2.4
Neth.	3.5	3.7	1.9	0.2	2.7	2.4
Norw.	3.6	3.3	4.4	2.3	3.6	3.5
Port.	6.5	7.7	1.5	1.1	4.9	4.2
Sp.	6.4	5.8	1.4	0.7	4.2	3.7
Sw.	3.6	3.1	1.5	1.7	2.6	2.5
Switz.	2.7	3.4	–0.1	1.4	2.1	1.8
Turk.	3.2	2.9	4.3	1.8	2.9	3.1
UK	2.4	2.9	1.5	1.1	1.8	1.9
Can.	3.5	4.1	2.9	1.3	4.0	3.0
USA	3.1	2.1	1.4	1.4	2.2	2.1
Jap.	9.3	6.8	2.5	3.3	6.4	5.7
Austral.	3.0	3.6	1.5	1.5	2.5	2.4
NZ	1.2	3.5	–0.7	2.0	1.6	1.4

Sources: (1)–(4), (6): OECD, *Economic Outlook: Historical Statistics* (1987): table 3.2; (5): OECD, *Economic Outlook: Historical Statistics* (1983): table 3.2.

TABLE 4.14. Growth of real GDP 1950–1981 (year to year averages)

	1950–60 (1)	1960–70 (2)	1970–81 (3)
Aus.	5.9	4.6	3.5
Belg.	2.7[a]	4.7	3.0
Den.	3.2	4.7	2.1
Fin.	4.4	4.3	3.1
Fr.	4.4	5.5	3.3
FRG	7.7	4.4	2.6
Gr.	6.0	7.5	4.3
Ice.	5.5	4.5	4.3
Ire.	1.3	4.2	4.0
It.	5.5[b]	5.7	2.9
Lux.	2.7[c]	3.4	2.4
Neth.	4.6	5.3	2.7
Norw.	3.6	4.9	4.5
Port.	4.1	5.9	4.2
Sp.	3.6[d]	7.1	3.4
Sw.	3.6	4.4	1.8
Switz.	4.2	4.3	0.7
Turk.	6.3	6.0	5.2
UK	2.7	2.9	1.7
Can.	4.0	5.6	3.8
USA	2.9	4.4	2.9
Jap.	8.0[e]	10.4	4.5
Austral.	4.3[a]	5.9	2.8
NZ	2.4	3.7	2.0

[a] 1953–60.
[b] 1951–60.
[c] 1950–9.
[d] 1954–60.
[e] 1952–60.

Sources: (1): UN, *Statistical Yearbook* (1970): table 181; (2): UN, *Statistical Yearbook* (1981): table 30; (3): World Bank (1984): Series 4, table 1.

TABLE 4.15. Growth of real GDP per capita 1950–1981 (year to year averages)

	1950–60 (1)	1960–70 (2)	1970–81 (3)
Aus.	5.7	4.0	3.3
Belg.	2.0[a]	4.1	2.8
Den.	2.5	3.9	1.8
Fin.	3.3	3.8	2.7
Fr.	3.5	4.4	2.8
FRG	6.6	3.5	2.6
Gr.	5.0	6.8	3.3
Ice.	3.2	2.9	3.1
Ire.	1.8	3.8	2.7
It.	4.9[b]	5.0	2.5
Lux.	2.0[c]	2.6	1.9
Neth.	3.3	4.0	1.9
Norw.	2.7	4.0	4.0
Port.	3.7	5.9	3.4
Sp.	2.8[d]	5.9	2.3
Sw.	2.9	3.6	1.5
Switz.	2.9	2.8	0.7
Turk.	3.4	3.4	2.8
UK	2.3	2.3	1.6
Can.	1.2	3.7	2.6
USA	1.2	3.1	1.9
Jap.	6.8[e]	9.3	3.4
Austral.	2.0[a]	3.7	1.4
NZ	0.1	1.9	0.5

[a] 1953–60.
[b] 1951–60.
[c] 1950–9.
[d] 1954–60.
[e] 1952–60.

Sources: (1): UN, *Statistical Yearbook* (1970): table 181; (2): UN, *Statistical Yearbook* (1981): table 30; (3): World Bank (1984): Series 4, table 1.

TABLE 4.16. Growth of GNP per capita 1960–1981
(year to year averages)

	1960–70 (1)	1970–80 (2)	1960–80 (3)	1965–84 (4)	1973–86 (5)
Aus.	3.9	3.4	4.1	3.5	2.4
Belg.	4.0	2.9	3.8	2.8	1.3
Den.	3.7	1.7	3.3	1.8	1.4
Fin.	3.9	2.5	4.0	3.3	2.4
Fr.	4.6	3.0	3.9	2.8	1.5
FRG	3.5	2.7	3.3	2.7	2.1
Gr.	6.6	3.7	5.8	3.6	1.4
Ice.	—	—	—	—	1.0
Ire.	3.6	2.6	3.1	2.2	1.4
It.	4.6	2.5	3.6	2.6	1.7
Lux.	—	—	—	—	2.7
Neth.	3.9	2.1	3.2	2.0	0.8
Norw.	4.1	3.8	3.5	3.3	3.4
Port.	5.3	1.2	5.0	3.3	1.1
Sp.	6.1	2.6	4.5	2.6	0.8
Sw.	3.8	1.6	2.3	1.8	1.1
Switz.	2.5	0.6	1.9	1.4	1.2
Turk.	3.9	3.0	3.6	2.6	1.4
UK	2.2	1.8	2.2	1.6	1.2
Can.	3.6	2.6	3.3	2.4	1.2
USA	3.2	2.1	2.3	1.7	1.4
Jap.	9.6	3.4	7.1	4.7	3.4
Austral.	3.1	1.3	2.7	2.0	1.4
NZ	2.1	0.3	1.8	1.4	0.7

Sources: (1): World Bank, *The World Bank Atlas 1972* (1972); (2): World Bank, *The World Bank Atlas 1983* (1983); (3): World Bank, *World Development Report 1982* (1982): table 1; (4): World Bank, *World Development Report 1987* (1987): table 1; (5): World Bank, *The 1988 Update of World Bank Atlas* (1988).

Section 5: Public Finance

TABLES

5.1. General government: current receipts as a percentage of GDP
5.2. General government: taxes as a percentage of GDP
5.3. General government: social security contributions as a percentage of GDP
5.4. General government: government final consumption as a percentage of GDP
5.5. General government: social security transfers as a percentage of GDP
5.6. General government: current disbursements as a percentage of GDP
5.7. Central government: current receipts as a percentage of GDP
5.8. Central government: taxes as a percentage of GDP
5.9. Central government: current disbursements as a percentage of GDP
5.10. Central government: government final consumption as a percentage of GDP
5.11. Central government: transfers to other sectors of general government as a percentage of central government disbursements
5.12. Central government: transfers to other sectors of general government as a percentage of GDP
5.13. Central government: social security transfers as a percentage of GDP
5.14. State and local government: taxes as a percentage of GDP
5.15. State and local government: final consumption as a percentage of GDP
5.16. Central government: taxes as a percentage of general government taxes
5.17. Central government: current receipts as a percentage of general government current receipts
5.18. General government: social security contributions as a percentage of total tax revenue
5.19. General government: property taxes as a percentage of total tax revenue

5.20. General government: taxes on payroll and work-force as a percentage of total tax revenue
5.21. General government: taxes on goods and services as a percentage of total tax revenue
5.22. General government: social security contributions as a percentage of GDP
5.23. General government: social security expenditure as a percentage of GDP
5.24. General government: health expenditure as a percentage of total government expenditure
5.25. General government: health expenditure as a percentage of GDP
5.26. General government: total health expenditure as a percentage of GDP
5.27. Central government: final consumption as a percentage of general government final consumption
5.28. Central government: current disbursements as a percentage of general government current disbursements
5.29. Central government: current disbursements minus transfers to other levels of general government as a percentage of general government current disbursements
5.30. General government: expenditure by function as a percentage of GDP 1975 and 1980
5.31. General government: total social expenditure as a percentage of total government expenditure
5.32. General government: total social expenditure as a percentage of GDP
5.33. General government: educational expenditure as a percentage of total government expenditure
5.34. General government: educational expenditure as a percentage of GDP
5.35. Central government: expenditure by function as a percentage of total expenditure, 1975
5.36. General government: military expenditure as a percentage of GDP
5.37. General government: official development assistance as a percentage of donor GNP
5.38. General government: total tax revenues as a percentage of GDP
5.39. General government: taxes on income and profits as a percentage of total tax revenue
5.40. General government: deficit as a percentage of GDP
5.41. Central government: deficit as a percentage of GDP

TABLE 5.1. General government: current receipts as a percentage of GDP

	1950	1955	1960	1965	1970	1975	1980	1985
Aus.	27.9	29.5	31.4	36.1	39.7	42.9	46.0	47.7
Belg.	24.2	24.0	26.7	30.7	35.2	40.4	43.2	46.5
Den.	21.7	25.7	27.6	31.4	41.7	46.2	52.2	57.0
Fin.	30.3	30.2	31.6	33.5	34.9	38.8	37.8	40.5
Fr.	32.6	33.0	34.1	37.7	39.0	40.3	45.6	48.5
FRG	31.6	34.7	36.0	36.2	38.5	42.7	44.6	45.4
Gr.	15.5	18.2	20.4	23.4	26.8	27.4	30.5	34.9
Ice.	27.6	26.9	36.4	29.0	31.8	35.6	32.1	33.4
Ire.	23.4	23.8	24.6	28.0	35.3	35.2	41.7	44.3
It.	21.0	26.2	29.8	31.6	30.4	31.2	37.4	37.5
Lux.	31.7	30.0	32.5	35.2	35.0	49.0	51.5	53.0
Neth.	33.0	28.9	33.4	36.8	44.5	53.2	55.0	54.4
Norw.	29.6	30.8	34.5	37.7	43.5	49.6	54.0	56.1
Port.	20.0	19.1	17.6	20.4	24.3	24.8	31.5	35.9
Sp.	—	—	18.1	19.3	22.5	24.4	30.0	33.2
Sw.	26.2	32.7	35.0	42.0	47.0	50.7	56.7	59.4
Switz.	25.5	24.0	25.4	25.4	26.5	32.1	32.8	34.4
Turk.	—	—	—	19.9	23.7	—	—	—
UK	33.5	30.4	29.6	32.8	40.7	40.8	40.9	43.7
Can.	24.1	26.0	28.0	30.6	35.2	36.9	37.2	38.8
USA	24.0	25.0	27.5	27.3	30.3	30.5	32.8	31.1
Jap.	21.9	19.9	20.7	20.8	20.7	24.0	28.0	31.2
Austral.	—	—	25.4	27.3	27.8	31.0	33.4	33.7
NZ	—	—	—	—	—	—	—	—

Note: Current receipts consist mainly of direct and indirect taxes, and social security contributions paid by employers and employees. General government consists of all departments, offices, organizations, and other bodies which are agents or instruments of the central, state, or local public authorities.

Sources: 1950–65: OECD, *National Accounts 1950–1968* (1968); OECD, *National Accounts 1960–1977* (1979); 1970–80: OECD, *National Accounts 1964–1981* (1983); 1985: OECD, *National Accounts 1973–1985* (1987); OECD, *National Accounts 1975–1987* (1989).

TABLE 5.2. General government: taxes as a percentage of GDP

	1950	1955	1960	1965	1970	1975	1980	1985
Aus.	22.4	23.6	24.4	27.5	27.4	28.6	29.2	30.7
Belg.	17.9	17.4	19.2	20.8	23.9	27.5	29.9	30.7
Den.	15.5	22.5	23.8	27.4	38.2	40.6	44.4	47.1
Fin.	25.2	24.0	24.9	26.1	27.0	29.5	29.5	30.9
Fr.	20.8	21.5	22.2	23.1	22.4	21.3	23.3	24.4
FRG	22.4	23.8	24.2	24.4	24.0	24.8	25.7	25.0
Gr.	12.3	13.2	14.1	16.1	18.0	18.0	19.0	20.9
Ice.	25.0	24.1	33.8	26.2	28.3	33.6	29.7	30.2
Ire.	20.3	20.5	21.2	24.1	28.4	26.7	30.4	32.8
It.	14.2	16.8	18.4	18.8	16.7	15.4	21.2	22.3
Lux.	24.1	21.1	21.4	21.8	21.7	29.5	30.8	31.8
Neth.	26.1	22.0	22.3	22.8	25.2	27.8	27.9	23.7
Norw.	26.1	26.3	26.6	27.8	31.5	33.8	38.6	37.9
Port.	14.4	13.3	13.5	14.9	18.0	16.4	20.8	24.0
Sp.	—	—	13.1	11.1	11.5	11.1	13.3	17.3
Sw.	21.9	26.6	27.0	31.5	33.2	35.2	35.3	37.5
Switz.	16.1	14.4	15.9	16.8	18.2	20.9	20.7	21.2
Turk.	—	—	—	15.0	17.3	—	—	—
UK	29.7	25.7	23.8	25.2	31.5	29.9	30.3	31.4
Can.	20.0	21.6	22.6	25.0	28.2	28.6	27.5	27.6
USA	21.5	22.2	23.4	23.0	23.4	22.0	23.0	21.4
Jap.	18.6	16.0	15.8	15.6	15.3	16.1	18.5	20.1
Austral.	—	—	22.3	23.5	24.3	28.4	30.2	30.2
NZ	—	—	—	—	—	—	—	—

Note: Taxes refer to direct as well as indirect taxes. Direct taxes mean levies by public authorities at regular intervals, except social contributions, on income from employment, property, capital gains, or any other sources. Indirect taxes mean taxes assessed on producers in respect of the production, sale, purchase, or use of goods and services, which they charge to the expenses of production. General government consists of all departments, offices, organizations, and other bodies which are agents or instruments of the central, state, or local public authorities.

Sources: 1950–65: OECD, *National Accounts 1950–1968* (1968); OECD, *National Accounts 1960–1977* (1979); 1970–80: OECD, *National Accounts 1964–1981* (1983); 1985: OECD, *National Accounts 1973–1985* (1987); OECD, *National Accounts 1975–1987* (1989).

TABLE 5.3. General government: social security contributions as a percentage of GDP

	1950	1955	1960	1965	1970	1975	1980	1985
Aus.	5.4	5.7	6.1	7.7	8.8	10.3	12.5	12.2
Belg.	5.1	5.4	6.4	9.2	10.3	12.6	12.5	14.7
Den.	1.3	1.4	1.5	1.9	1.9	0.6	0.9	1.9
Fin.	2.6	2.6	2.6	3.6	4.4	6.1	4.8	5.4
Fr.	9.3	10.8	11.2	13.8	12.9	15.3	18.4	19.9
FRG	7.7	8.2	9.7	9.7	11.6	15.0	15.5	16.2
Gr.	2.5	3.5	4.5	5.4	6.6	6.7	9.0	11.5
Ice.	2.2	2.2	2.0	2.3	2.4	0.9	1.4	1.2
Ire.	1.0	1.0	1.1	1.7	4.3	6.4	7.6	7.8
It.	4.9	6.9	8.7	10.1	10.7	13.0	12.6	11.8
Lux.	6.9	8.6	9.0	10.4	9.5	13.3	14.0	13.1
Neth.	4.3	4.7	8.1	11.3	14.6	18.4	18.2	20.5
Norw.	1.9	2.7	5.5	6.8	9.7	13.4	12.1	11.5
Port.	3.6	3.6	2.7	3.5	4.6	7.2	8.9	9.4
Sp.	—	—	3.5	5.0	6.6	9.4	12.0	12.0
Sw.	0.7	2.1	3.8	5.8	7.6	8.6	14.3	12.6
Switz.	4.2	4.2	4.7	4.8	5.6	8.3	9.1	9.9
Turk.	—	—	—	2.3	3.1	—	—	—
UK	3.4	3.1	3.6	4.8	5.2	6.6	6.2	6.9
Can.	1.4	1.7	2.0	1.9	2.9	3.5	3.4	4.4
USA	2.4	2.8	4.1	4.3	5.9	7.2	7.9	7.2
Jap.	1.9	2.2	2.5	3.3	4.3	6.4	7.4	8.3
Austral.	—	—	0.0	0.0	0.0	0.0	0.0	0.6
NZ	—	—	—	—	—	—	—	—

Note: Social security contributions include all compulsory contributions as payments to institutions of general government providing for social security benefits. General government consists of all departments, offices, organizations, and other bodies which are agents or instruments of the central, state, or local public authorities.

Sources: 1950–65: OECD, *National Accounts 1950–1968* (1968); OECD, *National Accounts 1960–1977* (1979); 1970–80: OECD, *National Accounts 1964–1981* (1983); *1985:* OECD, *National Accounts 1973–1985* (1987); OECD, *National Accounts 1975–1987* (1989).

TABLE 5.4. General government: government final consumption as a percentage of GDP

	1950	1955	1960	1965	1970	1975	1980	1985
Aus.	11.2	11.8	12.9	13.5	14.7	17.2	17.8	18.7
Belg.	12.6	11.4	12.6	13.1	13.7	16.8	18.5	17.7
Den.	10.2	12.9	12.7	15.4	19.8	24.6	26.7	25.4
Fin.	11.6	11.5	12.6	14.5	14.7	17.5	18.6	20.3
Fr.	12.8	13.0	12.9	12.6	13.4	14.4	15.2	16.3
FRG	14.3	13.2	13.6	15.4	15.8	20.5	20.2	19.9
Gr.	11.6	11.0	11.4	11.8	12.6	15.2	16.3	20.3
Ice.	8.6	7.9	8.0	8.0	8.7	10.0	16.4	18.2
Ire.	11.8	12.2	12.2	13.4	14.7	19.0	21.5	19.2
It.	11.6	11.8	12.1	14.1	13.8	15.4	16.4	16.6
Lux.	12.1	12.7	10.1	11.0	9.9	14.9	16.5	16.6
Neth.	12.5	14.6	13.5	15.5	16.3	18.2	18.0	16.3
Norw.	11.1	12.7	14.0	16.1	16.9	19.3	18.9	18.6
Port.	11.2	11.7	10.9	12.3	14.2	15.4	14.7	15.5
Sp.	—	—	8.8	7.4	8.5	9.2	11.6	13.7
Sw.	14.0	16.8	17.1	18.7	21.6	23.8	28.9	27.4
Switz.	11.4	10.5	10.2	11.6	10.5	12.6	12.7	13.2
Turk.	—	—	—	12.4	12.9	—	—	—
UK	16.4	17.1	16.8	17.2	17.7	22.0	21.5	21.1
Can.	10.3	14.3	14.3	13.9	19.2	20.0	19.6	20.1
USA	12.1	17.0	17.9	18.2	19.2	18.9	18.1	18.3
Jap.	10.8	10.4	8.9	9.3	7.4	10.1	10.0	9.7
Austral.	—	—	9.7	11.6	12.4	15.7	17.0	19.2
NZ	—	—	—	—	—	—	—	—

Note: Government final consumption consists of expenditures on goods and services for public administration, defence, health, and education. It excludes all transfer payments. General government consists of all departments, offices, organizations, and other bodies which are agents or instruments of the central, state, or local public authorities.

Sources: 1950–65: OECD, National Accounts 1950–1968 (1968); OECD, National Accounts 1960–1977 (1979); 1970–80: OECD, National Accounts 1964–1981 (1983); 1985: OECD, National Accounts 1973–1985 (1987); OECD, National Accounts 1975–1987 (1989).

TABLE 5.5. General government: social security transfers as a percentage of GDP

	1950	1955	1960	1965	1970	1975	1980	1985
Aus.	7.8	9.4	10.0	12.3	15.4	16.9	19.2	20.1
Belg.	9.7	9.0	10.9	12.4	14.1	18.8	21.2	22.0
Den.	5.8	6.7	7.3	8.3	11.6	13.9	16.8	16.5
Fin.	4.5	5.0	5.8	7.0	8.3	9.6	9.9	11.6
Fr.	11.2	13.1	12.9	16.1	17.2	20.6	23.3	26.6
FRG	12.3	11.4	12.6	13.0	13.4	18.4	17.3	16.6
Gr.	5.7	4.9	5.8	7.5	8.0	7.4	9.3	15.0
Ice.	6.4	5.9	7.1	7.2	8.9	10.5	4.6	5.2
Ire.	4.7	5.6	6.2	6.9	10.8	14.9	15.8	18.2
It.	6.3	9.4	10.7	13.4	12.7	16.0	16.1	16.9
Lux.	7.7	10.4	11.6	13.6	14.0	19.9	22.8	23.2
Neth.	6.7	7.2	10.3	14.0	18.4	25.6	27.4	28.5
Norw.	4.8	6.4	8.3	9.7	12.3	13.6	14.7	14.8
Port.	2.5	2.6	3.0	3.4	3.1	8.5	10.7	10.9
Sp.	—	—	2.3	5.7	7.5	9.3	13.8	16.0
Sw.	6.3	7.4	8.2	10.0	11.6	15.1	18.9	19.3
Switz.	4.9	5.3	6.2	7.3	6.3	10.1	10.3	13.7
Turk.	—	—	—	1.6	1.7	—	—	—
UK	5.4	5.4	6.2	7.4	9.4	10.9	12.5	14.6
Can.	5.6	6.3	8.5	8.6	8.1	10.2	10.1	12.3
USA	4.9	3.8	5.1	5.2	7.9	11.5	10.0	11.0
Jap.	2.5	4.1	3.8	4.4	4.8	8.0	10.6	11.5
Austral.	—	—	5.7	5.8	5.4	8.8	9.0	10.8
NZ	—	—	—	—	—	—	—	—

Note: Social security transfers consist of social security benefits, social assistance grants, unfunded employee pension and welfare benefits, and transfers to private non-profit institutions serving households. General government consists of all departments, offices, organizations, and other bodies which are agents or instruments of the central, state, or local public authorities.

Sources: 1950–65: OECD, *National Accounts 1950–1968* (1968); OECD, *National Accounts 1960–1977* (1979); 1970–80: OECD, *National Accounts 1964–1981* (1983); 1985: OECD, *National Accounts 1973–1985* (1987); OECD, *National Accounts 1975–1987* (1989).

TABLE 5.6. General government: current disbursements as a percentage of GDP

	1950	1955	1960	1965	1970	1975	1980	1985
Aus.	21.2	23.0	25.4	28.9	33.1	38.6	42.7	45.2
Belg.	25.5	23.8	27.8	29.8	33.0	41.2	48.1	52.3
Den.	18.0	21.4	21.7	25.9	34.6	43.5	52.2	56.7
Fin.	19.7	20.7	21.9	25.8	28.9	32.2	34.3	37.7
Fr.	26.7	29.8	30.2	32.9	34.7	39.2	43.1	49.4
FRG	28.3	27.0	28.2	30.4	32.6	43.4	42.8	43.4
Gr.	19.6	16.3	17.8	21.3	22.4	26.7	30.4	45.3
Ice.	19.9	20.0	23.4	20.6	21.7	28.3	25.0	28.3
Ire.	22.9	23.4	24.5	27.6	34.2	42.0	48.3	50.4
It.	20.7	24.6	26.6	30.9	30.2	38.3	41.4	44.1
Lux.	22.5	26.9	25.5	29.7	28.6	41.3	45.7	47.8
Neth.	23.9	25.5	28.0	33.0	40.2	51.1	54.2	55.2
Norw.	21.9	24.4	28.0	31.9	36.5	41.8	45.1	44.0
Port.	16.3	15.9	15.2	17.7	19.5	27.2	33.8	39.4
Sp.	—	—	13.7	15.8	18.8	21.2	29.4	34.7
Sw.	23.5	26.4	28.7	31.9	37.2	44.9	57.1	60.8
Switz.	19.4	18.5	19.1	21.3	21.3	28.8	29.3	30.9
Turk.	—	—	—	15.5	16.4	—	—	—
UK	30.1	28.8	29.3	30.5	33.2	41.0	42.3	44.9
Can.	19.2	23.4	26.6	26.4	32.2	36.8	37.7	43.8
USA	20.0	22.5	25.0	25.2	30.3	33.6	33.5	35.3
Jap.	14.6	15.5	13.6	14.7	14.0	20.9	25.4	26.9
Austral.	—	—	18.9	21.6	21.8	27.6	30.4	35.5
NZ	—	—	—	—	—	—	—	—

Note: Current disbursements consist of final consumption expenditure, interest on the public debt, subsidies, and social security transfers to households. General government consists of all departments, offices, organizations, and other bodies which are agents or instruments of the central, state, or local public authorities.

Sources: 1950–65: OECD, *National Accounts 1950–1968* (1968); OECD, *National Accounts 1960–1977* (1979); 1970–80: OECD, *National Accounts 1964–1981* (1983); 1985: OECD, *National Accounts 1973–1985* (1987); OECD, *National Accounts 1975–1987* (1989).

TABLE 5.7. Central government: current receipts as a percentage of GDP

	1950	1955	1960	1965	1970	1975	1980	1985
Aus.	13.2	16.6	17.2	19.1	21.0	21.0	22.0	24.1
Belg.	17.5	16.7	18.0	19.2	22.1	24.9	27.8	28.4
Den.	—	—	19.1	22.1	—	—	36.0	40.6
Fin.	—	—	21.4	22.2	22.5	22.8	22.8	24.1
Fr.	—	—	21.0	21.9	22.0	19.6	21.7	21.5
FRG	11.0	15.6	13.8	14.0	14.3	13.9	14.5	14.7
Gr.	10.5	12.4	13.1	14.3	16.1	17.4	18.1	19.3
Ice.	19.7	18.7	28.0	19.5	24.0	27.6	25.4	26.1
Ire.	19.1	19.3	20.2	23.4	29.4	28.9	35.6	37.9
It.	13.2	15.9	17.6	17.8	17.0	18.9	26.9	28.5
Lux.	21.5	18.5	19.3	19.6	20.3	28.9	29.3	32.1
Neth.	—	22.3	22.6	23.1	27.0	32.0	33.8	30.1
Norw.	22.3	22.5	26.1	28.5	35.0	40.4	49.5	36.1
Port.	14.6	13.5	13.0	15.1	17.6	16.4	22.0	25.7
Sp.	—	—	—	12.5	13.9	13.2	15.6	16.2
Sw.	19.9	21.9	22.9	24.9	26.8	28.7	27.7	31.3
Switz.	—	—	8.6	8.3	8.6	8.4	9.2	9.6
Turk.	—	—	—	15.4	18.1	—	—	—
UK	27.9	25.0	23.1	24.2	31.5	29.9	30.7	32.3
Can.	15.1	16.5	16.0	15.8	17.5	18.7	16.7	17.4
USA	15.5	15.8	15.6	14.6	14.5	12.7	14.6	13.6
Jap.	—	—	14.2	11.5	11.0	10.6	13.1	13.5
Austral.	—	—	20.3	21.3	21.9	23.9	26.0	27.0
NZ	—	—	—	—	—	—	—	—

Note: Current receipts consist of direct and indirect taxes, and social security contributions paid by employers and employees. Central government consists of all departments, offices, organizations, and other bodies classified under general government which are agencies or instruments of the central authority of a country, except separately organized social security funds.

Sources: 1950–65: OECD, *National Accounts 1950–1968* (1968); OECD, *National Accounts 1960–1977* (1979); 1970–80: OECD, *National Accounts 1964–1981* (1983); 1985: OECD, *National Accounts 1973–1985* (1987); OECD, *National Accounts 1975–1987* (1989).

TABLE 5.8. Central government: taxes as a percentage of GDP

	1950	1955	1960	1965	1970	1975	1980	1985
Aus.	13.1	16.6	16.6	18.4	17.9	18.4	19.0	20.5
Belg.	17.1	16.5	17.9	19.5	22.3	25.7	28.2	28.5
Den.	—	—	17.9	21.2	—	—	30.6	33.5
Fin.	—	—	19.7	20.4	19.4	20.6	21.1	21.8
Fr.	—	—	20.4	21.2	20.2	18.4	20.2	20.1
FRG	9.9	14.1	13.0	13.3	13.4	13.0	13.3	12.8
Gr.	10.3	11.5	12.2	13.2	15.4	16.2	17.4	18.8
Ice.	19.5	18.5	27.7	19.6	21.5	26.4	23.4	23.6
Ire.	17.0	17.0	17.5	20.7	17.5	19.8	26.8	32.0
It.	12.0	14.2	15.9	16.2	14.4	14.6	20.3	21.1
Lux.	21.5	18.2	18.1	18.4	17.9	24.6	25.1	26.8
Neth.	—	21.5	21.7	22.2	24.8	27.4	27.6	22.7
Norw.	20.9	20.9	23.5	25.9	22.5	23.7	29.7	29.3
Port.	13.0	11.8	12.0	13.6	16.2	15.6	20.3	23.4
Sp.	—	—	—	9.7	9.8	9.5	10.9	12.3
Sw.	18.2	20.3	20.6	20.7	22.6	23.9	20.4	23.1
Switz.	—	—	1.8	1.4	8.3	7.9	8.6	8.7
Turk.	—	—	—	14.0	16.4	—	—	—
UK	27.1	23.2	20.7	21.8	28.0	23.3	27.2	27.5
Can.	13.9	15.4	14.6	14.2	14.9	15.6	13.7	13.5
USA	15.3	15.6	15.5	14.4	14.0	12.2	13.9	11.9
Jap.	—	—	12.9	11.0	10.7	10.1	12.3	12.8
Austral.	—	—	19.3	20.2	20.9	23.1	24.9	24.6
NZ	—	—	—	—	—	—	—	—

Note: Taxes refer to direct as well as indirect taxes. Central government consists of all departments, offices, organizations, and other bodies classified under general government which are agencies or instruments of the central authority of a country, except separately organized social security funds.

Sources: 1950–65: OECD, *National Accounts 1950–1968* (1968); OECD, *National Accounts 1960–1977* (1979); 1970–80: OECD, *National Accounts 1964–1981* (1983); 1985: OECD, *National Accounts 1973–1985* (1987); OECD, *National Accounts 1975–1987* (1989).

TABLE 5.9. Central government: current disbursements as a percentage of GDP

	1950	1955	1960	1965	1970	1975	1980	1985
Aus.	12.4	12.7	14.0	15.9	18.7	21.5	22.1	24.8
Belg.	18.8	16.9	20.0	19.5	21.3	26.5	31.7	35.1
Den.	—	—	14.4	17.8	—	—	38.7	42.3
Fin.	—	—	14.6	17.3	17.3	19.5	21.3	22.4
Fr.	—	—	18.2	18.2	19.0	19.9	20.8	23.6
FRG	10.2	11.7	10.8	12.0	11.5	14.5	14.1	14.1
Gr.	16.1	11.7	11.8	14.0	14.3	19.4	20.3	27.6
Ice.	13.4	14.1	17.7	14.4	16.8	24.9	20.6	22.9
Ire.	18.8	18.8	20.3	23.0	28.0	34.4	41.2	43.4
It.	13.4	15.3	14.9	17.3	16.7	21.4	31.5	36.0
Lux.	15.7	16.8	15.3	16.8	16.4	23.2	25.4	28.1
Neth.	—	20.1	18.5	19.7	23.6	30.0	33.1	31.7
Norw.	16.5	18.3	21.5	24.7	30.5	34.5	41.9	38.4
Port.	12.8	12.4	11.8	13.8	15.5	19.0	24.6	30.5
Sp.	—	—	—	9.5	11.4	10.8	15.3	18.4
Sw.	17.9	18.6	17.8	20.6	22.9	27.3	33.0	35.8
Switz.	—	—	5.3	5.7	6.9	7.7	8.7	9.1
Turk.	—	—	—	14.0	13.3	—	—	—
UK	25.6	23.4	22.9	22.1	24.1	31.2	31.9	33.5
Can.	11.9	16.0	16.0	14.3	17.0	20.4	19.8	23.4
USA	12.3	14.8	15.0	14.6	16.5	16.4	16.3	18.4
Jap.	—	—	5.9	5.9	8.7	11.7	15.1	15.0
Austral.	—	—	15.3	17.1	17.3	22.7	24.4	27.5
NZ	—	—	—	—	—	—	—	—

Note: Current disbursements consist of final consumption expenditure, interest on the public debt, subsidies, and social security transfers to households. Central government consists of all departments, offices, organizations, and other bodies classified under general government which are agencies or instruments of the central authority of a country, except separately organized social security funds.

Sources: 1950–65: OECD, *National Accounts 1950–1968* (1968); OECD, *National Accounts 1960–1977* (1979); 1970–80: OECD, *National Accounts 1964–1981*(1983); 1985: OECD, *National Accounts 1973–1985* (1987); OECD, *National Accounts 1975–1987* (1989).

TABLE 5.10. Central government: government final consumption as a percentage of GDP

	1950	1955	1960	1965	1970	1975	1980	1985
Aus.	5.9	5.5	6.3	6.4	6.8	6.4	6.3	7.0
Belg.	9.6	8.1	9.8	10.1	10.6	12.9	13.7	13.0
Den.	—	—	5.8	7.0	—	—	8.3	7.9
Fin.	—	—	5.5	6.4	5.9	6.3	6.1	6.2
Fr.	—	—	10.0	9.5	10.2	10.6	10.9	11.4
FRG	4.3	4.3	4.1	5.0	3.9	4.3	3.9	3.8
Gr.	9.7	9.3	9.0	9.0	9.8	12.4	12.1	14.4
Ice.	7.6	5.3	5.3	5.0	5.6	6.8	9.0	10.1
Ire.	6.7	6.8	7.0	7.8	7.4	9.3	9.9	9.5
It.	8.4	8.3	8.2	9.7	7.9	8.5	9.0	8.8
Lux.	8.9	9.1	7.2	7.6	6.8	10.1	10.9	10.9
Neth.	—	8.0	6.3	6.5	7.8	8.2	8.1	7.5
Norw.	6.2	7.1	7.2	8.0	7.9	8.2	7.5	7.0
Port.	9.8	10.4	9.5	11.1	12.5	12.2	12.8	13.6
Sp.	—	—	—	5.8	6.6	6.5	7.8	6.9
Sw.	8.2	8.9	7.8	9.2	8.1	8.1	8.7	7.2
Switz.	—	—	2.8	3.2	2.9	2.9	3.0	3.3
Turk.	—	—	—	10.6	10.6	—	—	—
UK	11.6	12.3	11.3	10.8	10.5	12.4	12.7	12.9
Can.	5.2	8.7	6.6	5.4	5.2	5.0	4.6	5.0
USA	6.3	10.8	10.4	9.8	9.9	8.1	7.7	8.7
Jap.	—	—	3.8	3.7	2.1	2.4	2.4	2.4
Austral.	—	—	4.4	5.9	5.8	5.6	5.7	6.3
NZ	—	—	—	—	—	—	—	—

Note: Government final consumption consists of goods and services for public administration, defence, health, and education. It excludes all transfer payments. Central government consists of all departments, offices, organizations, and other bodies classified under general government which are agencies or instruments of the central authority of a country, except separately organized social security funds.

Sources: 1950–65: OECD, *National Accounts 1950–1968* (1968); OECD, *National Accounts 1960–1977* (1979); 1970–80: OECD, *National Accounts 1964–1981* (1983); 1985: OECD, *National Accounts 1973–1985* (1987); OECD, *National Accounts 1975–1987* (1989).

TABLE 5.11. Central government: transfers to other sectors of general government as a percentage of central government disbursements

	1950	1955	1960	1965	1970	1975	1980	1985
Aus.	4.5	8.0	10.3	16.2	18.3	26.9	21.7	24.0
Belg.	19.8	20.3	21.8	20.9	21.6	26.6	28.0	25.1
Den.	—	—	41.5	40.1	—	—	55.6	45.1
Fin.	—	—	18.9	20.2	25.2	27.4	30.2	34.2
Fr.	—	—	7.0	8.5	12.3	13.8	15.8	15.4
FRG	15.3	24.2	36.5	32.4	25.3	28.3	28.2	24.9
Gr.	4.0	5.5	6.7	11.9	14.2	11.9	13.1	14.5
Ice.	10.3	11.4	18.4	21.2	39.5	35.6	34.0	32.0
Ire.	17.2	18.9	19.8	21.7	22.6	26.5	30.2	27.9
It.	10.0	13.4	13.8	21.9	19.8	27.4	41.4	42.7
Lux.	21.4	17.5	21.1	21.3	22.8	19.7	20.1	21.1
Neth.	—	35.0	45.9	51.0	47.1	50.0	51.4	48.1
Norw.	12.9	15.1	9.5	10.4	10.9	10.4	26.6	25.5
Port.	0.7	0.6	0.8	0.2	0.6	5.6	5.8	8.3
Sp.	—	—	—	1.8	4.9	5.4	13.1	26.4
Sw.	9.4	11.5	30.1	32.5	23.1	27.7	23.1	22.7
Switz.	—	—	10.3	16.5	36.7	40.0	41.6	40.2
Turk.	—	—	—	10.2	5.1	—	—	—
UK	14.1	13.1	16.6	19.8	23.5	27.2	22.6	19.9
Can.	11.5	10.3	16.9	18.9	23.1	22.5	21.7	19.7
USA	6.6	5.3	8.5	11.0	17.9	24.4	23.5	15.3
Jap.	—	—	—	—	54.2	57.3	53.1	49.3
Austral.	—	—	26.9	25.3	28.6	33.4	32.4	27.0
NZ	—	—	—	—	—	—	—	—

Note: Central government consists of all departments, offices, organizations, and other bodies classified under general government which are agencies or instruments of the central authority of a country, except separately organized social security funds.

Sources: 1950–65: OECD, *National Accounts 1950–1968* (1968); OECD, *National Accounts 1960–1977* (1979); 1970–80: OECD, *National Accounts 1964–1981* (1983); 1985: OECD, *National Accounts 1973–1985* (1987); OECD, *National Accounts 1975–1987* (1989).

TABLE 5.12. Central government: transfers to other sectors of general government as a percentage of GDP

	1950	1955	1960	1965	1970	1975	1980	1985
Aus.	0.6	1.0	1.4	2.6	3.4	5.8	4.8	6.0
Belg.	3.7	3.4	4.4	4.1	4.6	7.1	8.9	8.8
Den.	—	—	6.0	7.1	—	—	21.5	19.1
Fin.	—	—	2.8	3.5	4.4	5.4	6.4	7.7
Fr.	—	—	1.3	1.6	2.3	2.7	3.3	3.6
FRG	1.6	2.8	4.0	3.9	2.9	4.1	4.0	3.5
Gr.	0.6	0.6	0.8	1.7	2.0	2.3	2.7	4.0
Ice.	1.4	1.6	3.3	3.0	6.6	8.9	7.0	7.3
Ire.	3.2	3.6	4.0	5.0	6.3	9.1	12.4	12.1
It.	1.3	2.1	2.1	3.8	3.3	5.9	13.1	15.4
Lux.	3.4	2.9	3.2	3.6	3.7	4.6	5.1	5.9
Neth.	—	7.0	8.5	10.0	11.1	15.0	17.0	15.2
Norw.	2.1	2.8	2.1	2.6	3.3	3.6	11.1	9.8
Port.	0.1	0.1	0.1	0.0	0.1	1.1	1.4	2.5
Sp.	—	—	—	0.2	0.6	0.6	2.0	4.9
Sw.	1.7	2.2	5.4	6.7	5.3	7.5	7.6	8.1
Switz.	—	—	0.6	1.0	2.5	3.1	3.6	3.6
Turk.	—	—	—	1.4	0.7	—	—	—
UK	3.6	3.1	3.8	4.4	5.7	8.5	7.2	6.7
Can.	1.4	1.6	2.7	2.7	3.9	4.6	4.3	4.6
USA	0.8	0.8	1.3	1.6	3.0	4.0	3.8	2.8
Jap.	—	—	—	—	4.7	6.7	8.0	7.4
Austral.	—	—	4.1	4.3	5.0	7.6	7.9	7.4
NZ	—	—	—	—	—	—	—	—

Note: Central government consists of all departments, offices, organizations, and other bodies classified under general government which are agencies or instruments of the central authority of a country, except separately organized social security funds.

Sources: 1950–65: OECD, *National Accounts 1950–1968* (1968); OECD, *National Accounts 1960–1977* (1979); 1970–80: OECD, *National Accounts 1964–1981* (1983); 1985: OECD, *National Accounts 1973–1985* (1987); OECD, *National Accounts 1975–1987* (1989).

TABLE 5.13. Central government: social security transfers as a percentage of GDP

	1950	1955	1960	1965	1970	1975	1980	1985
Aus.	4.0	4.5	4.1	4.4	6.0	6.4	7.0	6.9
Belg.	2.6	2.2	2.0	1.4	1.5	1.8	2.0	2.0
Den.	—	—	1.4	2.0	—	—	1.3	1.5
Fin.	—	—	3.0	3.3	3.3	3.6	3.9	3.5
Fr.	—	—	2.9	3.5	3.2	3.3	3.3	3.9
FRG	3.3	2.9	2.0	2.2	2.2	3.0	2.5	2.3
Gr.	3.5	1.4	1.3	1.4	0.6	0.6	0.7	1.0
Ice.	1.7	1.5	1.2	1.2	0.9	1.5	1.0	1.0
Ire.	3.3	4.0	4.1	3.8	6.2	8.1	8.0	9.0
It.	1.1	2.0	1.6	1.3	2.4	1.7	2.2	2.8
Lux.	0.9	1.2	1.4	0.9	1.4	2.2	3.0	3.4
Neth.	—	2.6	1.2	1.2	1.7	2.5	2.4	2.6
Norw.	2.4	3.4	7.0	8.6	13.0	14.8	13.1	12.2
Port.	0.4	0.4	0.6	0.5	0.5	2.3	2.2	1.5
Sp.	—	—	—	1.2	1.8	1.3	1.8	2.4
Sw.	5.2	5.5	1.9	2.5	7.1	7.8	9.8	8.8
Switz.	—	—	0.4	0.3	0.0	0.0	0.0	0.3
Turk.	—	—	—	1.1	1.2	—	—	—
UK	2.3	2.1	2.1	2.1	3.4	4.0	5.2	6.5
Can.	2.7	3.5	3.9	3.6	4.7	6.4	5.5	6.7
USA	1.5	0.9	0.8	0.8	1.0	1.9	1.6	2.1
Jap.	—	—	1.4	1.3	0.5	0.6	0.7	0.8
Austral.	—	—	5.6	5.6	5.1	8.5	8.7	9.7
NZ	—	—	—	—	—	—	—	—

Note: Social security transfers consist of social security benefits, social assistance grants, unfunded employee pension and welfare benefits, and transfers to private non-profit institutions serving households. Central government consists of all departments, offices, organizations, and other bodies classified under general government which are agencies or instruments of the central authority of a country, except separately organized social security funds.

Sources: 1950–65: OECD, *National Accounts 1950–1968* (1968); OECD, *National Accounts 1960–1977* (1979); 1970–80: OECD, *National Accounts 1964–1981* (1983); 1985: OECD, *National Accounts 1973–1985* (1987); OECD, *National Accounts 1975–1987* (1989).

TABLE 5.14. State and local government: taxes as a percentage of GDP

	1975	1980	1985
Aus.	10.5	10.2	10.2
Belg.	1.8	1.7	2.2
Den.	—	13.8	13.8
Fin.	8.7	8.1	9.0
Fr.	2.8	3.0	4.0
FRG	11.7	12.4	12.2
Gr.	1.0	0.8	0.8
Ice.	—	6.3	6.6
Ire.	2.8	1.8	1.5
It.	0.9	1.0	1.4
Lux.	4.9	5.8	5.0
Neth.	0.5	0.9	1.1
Norw.	10.0	8.9	8.6
Port.	—	0.7	0.6
Sp.	—	2.4	4.9
Sw.	11.3	14.8	14.3
Switz.	5.9	5.2	5.5
Turk.	—	—	—
UK	3.8	3.6	3.9
Can.	12.6	15.8	14.1
USA	9.5	8.9	9.5
Jap.	5.8	6.8	7.5
Austral.	5.2	5.3	5.7
NZ	—	—	—

Note: Taxes refer to direct as well as indirect taxes. State and local government includes all kinds of subcentral governments.

Source: OECD, *National Accounts 1973–1985* (1987).

TABLE 5.15. State and local government: final consumption as a percentage of GDP

	1975	1980	1985
Aus.	7.1	7.5	7.6
Belg.	3.0	3.7	3.6
Den.	—	18.2	17.2
Fin.	10.6	11.7	13.4
Fr.	2.8	3.3	3.8
FRG	9.9	10.1	9.7
Gr.	2.0	3.8	4.1
Ice.	—	3.8	4.0
Ire.	9.2	10.5	9.4
It.	6.2	6.7	8.6
Lux.	4.0	4.6	4.6
Neth.	9.0	9.2	8.0
Norw.	11.0	11.3	11.4
Port.	—	1.1	1.4
Sp.	—	2.2	4.2
Sw.	15.5	19.8	19.9
Switz.	4.1	4.0	4.0
Turk.	—	—	—
UK	9.1	8.2	8.0
Can.	13.3	14.7	15.0
USA	10.5	9.9	9.6
Jap.	7.8	7.5	7.3
Austral.	11.9	12.8	12.9
NZ	—	—	—

Note: Government final consumption consists of goods and services for public administration, defence, health, and education. It excludes all transfer payments. State and local government includes all kinds of sub-central governments.

Source: 1970–85: OECD, *National Accounts 1973–1985* (1987).

TABLE 5.16. Central government: taxes as a percentage of general government taxes

	1950	1955	1960	1965	1970	1975	1980	1985
Aus.	58.5	70.3	67.9	67.1	65.4	64.4	65.0	66.8
Belg.	95.4	94.6	93.4	93.8	93.2	93.5	94.3	92.7
Den.	—	—	75.2	77.3	—	—	69.0	71.0
Fin.	—	—	79.0	78.2	72.0	69.9	71.7	70.7
Fr.	—	—	91.8	91.7	90.1	86.3	86.7	82.6
FRG	44.1	59.2	53.8	54.7	55.9	52.6	51.7	51.2
Gr.	84.1	87.2	86.1	82.2	85.5	89.6	91.6	90.1
Ice.	78.2	76.8	82.0	74.7	76.0	78.8	78.9	78.2
Ire.	83.8	82.8	82.5	86.2	61.4	74.0	88.0	97.4
It.	84.1	85.0	86.5	86.1	86.5	94.7	95.7	95.0
Lux.	88.9	86.5	84.9	84.5	82.9	83.3	81.5	84.2
Neth.	—	97.9	97.6	97.5	98.2	98.6	98.8	95.5
Norw.	80.3	79.7	88.5	93.0	71.5	70.3	76.9	77.3
Port.	90.2	89.1	89.5	91.2	90.1	94.6	97.5	97.3
Sp.	—	—	—	87.6	85.6	85.3	81.9	71.1
Sw.	83.4	76.4	76.1	65.8	68.0	67.9	57.8	61.6
Switz.	—	—	11.0	8.3	45.4	37.8	41.4	40.9
Turk.	—	—	—	93.4	94.7	—	—	—
UK	91.2	90.3	87.3	86.2	88.7	77.8	89.6	87.7
Can.	69.5	71.3	64.7	56.9	52.9	54.6	49.6	48.8
USA	71.1	70.3	66.2	62.5	59.9	55.6	60.3	55.5
Jap.	—	—	81.4	70.2	69.8	62.5	66.5	63.6
Austral.	—	—	86.4	85.7	86.1	81.5	82.5	81.3
NZ	—	—	—	—	—	—	—	—

Note: Central government consists of all departments, offices, organizations, and other bodies classified under general government which are agencies or instruments of the central authority of a country, except separately organized social security funds.

Sources: 1950–65: OECD, *National Accounts 1950–1968* (1968); OECD, *National Accounts 1960–1977* (1979); 1970–80: OECD, *National Accounts 1964–1981* (1983); 1985: OECD, *National Accounts 1973–1985* (1987); OECD, *National Accounts, 1965–1987* (1989).

TABLE 5.17. Central government: current receipts as a percentage of general government current receipts

	1950	1955	1960	1965	1970	1975	1980	1985
Aus.	47.3	56.2	54.8	52.9	52.8	49.0	47.8	50.5
Belg.	72.0	69.6	67.7	62.6	62.6	61.5	64.3	61.1
Den.	—	—	69.5	70.4	—	—	69.0	71.3
Fin.	—	—	67.5	66.0	64.4	58.9	60.4	59.5
Fr.	—	—	61.5	58.0	56.3	48.7	47.5	44.2
FRG	35.0	44.9	38.3	38.7	37.3	32.5	32.5	32.4
Gr.	67.9	68.0	64.3	61.0	60.1	63.4	59.4	55.4
Ice.	71.6	69.8	76.8	67.3	75.4	77.8	79.1	78.1
Ire.	81.4	81.1	82.0	83.8	83.4	81.7	85.3	85.6
It.	62.8	60.6	59.1	56.4	56.0	60.4	71.8	76.0
Lux.	67.7	61.6	59.4	55.6	58.1	58.7	57.0	60.5
Neth.	—	77.3	67.7	62.7	60.7	60.2	61.5	55.3
Norw.	75.5	73.2	75.5	75.6	80.3	81.3	91.7	64.4
Port.	72.8	71.0	74.2	73.9	72.4	66.3	70.0	71.5
Sp.	—	—	—	65.1	61.8	54.1	51.9	48.8
Sw.	76.1	67.0	65.6	59.2	56.9	56.7	48.8	52.6
Switz.	—	—	33.9	32.8	32.6	26.3	27.9	27.8
Turk.	—	—	—	77.3	76.5	—	—	—
UK	83.0	82.2	77.8	73.9	77.3	73.4	75.0	73.9
Can.	62.5	63.6	57.1	51.5	49.9	50.8	44.9	44.9
USA	64.6	63.0	56.9	53.3	47.7	41.7	44.4	43.6
Jap.	—	—	68.7	55.4	53.1	43.9	46.7	43.3
Austral.	—	—	79.9	78.0	79.0	77.2	77.8	80.2
NZ	—	—	—	—	—	—	—	—

Note: Central government consists of all departments, offices, organizations, and other bodies classified under general government which are agencies or instruments of the central authority of a country, except separately organized social security funds.

Sources: 1950–65: OECD, *National Accounts 1950–1968* (1968); OECD, *National Accounts 1960–1977* (1979); 1970–80: OECD, *National Accounts 1964–1981* (1983); 1985: OECD, *National Accounts 1973–1985* (1987); OECD, *National Accounts 1975–1987* (1989).

TABLE 5.18. General government: social security contributions as a percentage of total tax revenue

	1955	1960	1965	1970	1975	1980	1985
Aus.	23.8	24.2	24.9	25.4	27.7	31.3	31.8
Belg.	25.9	27.0	31.4	30.5	31.9	30.4	33.1
Den.	4.7	5.0	5.4	4.0	1.3	1.8	3.8
Fin.	7.9	6.8	2.9	4.8	8.3	8.2	9.1
Fr.	—	—	34.2	36.3	40.8	43.1	43.3
FRG	24.5	27.6	26.8	30.3	34.1	34.4	36.5
Gr.	—	—	26.9	27.1	27.2	33.6	35.7
Ice.	—	—	—	—	—	—	—
Ire.	4.6	4.9	6.5	8.2	13.8	14.3	14.8
It.	32.1	32.7	34.2	37.8	45.9	36.6	34.7
Lux.	—	—	32.3	28.6	30.3	29.2	25.5
Neth.	16.1	25.4	30.6	35.1	38.4	38.1	44.3
Norw.	4.5	8.8	12.0	16.1	24.9	21.0	20.6
Port.	17.2	18.3	21.9	23.9	34.6	29.5	26.0
Sp.	—	—	28.3	37.4	47.5	48.6	41.5
Sw.	2.1	4.3	12.1	14.9	19.5	28.6	24.8
Switz.	21.5	21.6	22.5	23.4	29.2	30.9	32.0
Turk.	5.3	7.1	5.9	6.3	9.5	5.6	14.3
UK	10.4	12.6	15.4	13.9	17.4	16.9	17.8
Can.	4.2	5.7	5.7	9.6	10.1	10.4	13.5
USA	11.0	14.4	16.4	19.3	24.5	26.1	29.4
Jap.	12.7	13.8	21.8	22.3	29.0	29.1	30.3
Austral.	0.0	0.0	0.0	0.0	0.0	0.0	0.0
NZ	22.4	20.5	0.0	0.0	0.0	0.0	0.0

Note: Social security contributions include all compulsory contributions as payments to institutions of general government providing for social security benefits. General government consists of all departments, offices, organizations, and other bodies which are agents or instruments of the central, state, or local public authorities.

Sources: OECD, *Revenue Statistics 1965–1983* (1984); OECD, *Revenue Statistics 1965–1987* (1988).

TABLE 5.19. General government: property taxes as a percentage of total tax revenue

	1955	1960	1965	1970	1975	1980	1985
Aus.	3.3	4.2	3.7	3.7	3.0	2.8	2.5
Belg.	3.9	4.1	3.7	3.0	2.4	2.4	1.7
Den.	10.1	10.7	8.0	6.0	5.9	5.7	4.3
Fin.	6.0	6.0	4.1	2.3	2.1	2.0	3.0
Fr.	—	—	4.3	3.5	3.4	3.6	4.5
FRG	9.8	7.1	5.8	4.9	3.9	3.3	3.0
Gr.	—	—	10.3	9.7	10.0	5.2	2.7
Ice.	—	—	—	—	—	—	—
Ire.	19.2	19.1	15.1	12.2	9.7	5.3	4.0
It.	8.1	7.9	7.2	6.0	3.3	3.9	2.5
Lux.	—	—	6.2	6.6	4.9	5.7	5.6
Neth.	7.0	5.6	4.9	3.3	2.6	3.9	3.5
Norw.	4.0	3.7	3.1	2.4	2.3	1.7	1.9
Port.	6.7	7.0	5.1	4.2	2.5	1.5	1.9
Sp.	—	—	6.4	6.5	6.3	4.6	2.8
Sw.	2.4	2.2	1.8	1.5	1.1	0.9	2.3
Switz.	8.9	8.5	8.8	8.8	7.1	7.3	8.2
Turk.	0.7	0.7	10.5	10.8	6.9	6.2	4.6
UK	12.9	15.2	14.5	12.4	12.7	12.0	12.0
Can.	11.3	15.6	13.2	13.0	9.4	9.1	9.4
USA	13.5	14.3	15.3	13.6	13.3	10.1	10.1
Jap.	9.9	9.3	8.1	7.6	9.1	8.2	9.7
Austral.	10.1	11.9	11.7	11.0	9.4	8.3	7.8
NZ	11.5	12.6	11.5	10.4	9.3	7.4	7.1

Note: Taxes on property cover current and non-recurrent taxes on the use, ownership, or transfer of property. General government consists of all departments, offices, organizations, and other bodies which are agents or instruments of the central, state, or local public authorities.

Sources OECD, *Revenue Statistics 1965–1983* (1984); OECD, *Revenue Statistics 1965–1987* (1988).

TABLE 5.20. General government: taxes on payroll and work-force as a percentage of total tax revenue

	1955	1960	1965	1970	1975	1980	1985
Aus.	8.3	8.3	7.6	7.7	8.0	7.0	5.7
Belg.	0.0	0.0	0.0	0.0	0.0	0.0	0.0
Den.	0.0	0.0	0.0	0.0	0.0	0.0	0.8
Fin.	0.0	0.0	5.3	4.6	2.5	0.2	0.5
Fr.	—	—	4.6	1.2	1.9	2.1	2.1
FRG	0.8	0.7	0.6	0.6	0.8	0.2	0.0
Gr.	—	—	0.8	0.7	0.8	0.2	1.5
Ice.	—	—	—	—	—	—	—
Ire.	0.0	0.0	0.0	0.0	0.0	0.0	0.0
It.	0.0	0.0	0.0	0.0	0.0	0.6	0.6
Lux.	—	—	0.9	1.0	0.9	0.7	0.6
Neth.	4.0	0.0	0.0	0.0	0.0	0.0	0.0
Norw.	0.0	0.0	0.0	0.0	0.0	0.0	0.0
Port.	0.8	0.8	0.9	1.0	2.5	2.6	2.5
Sp.	—	—	0.0	0.0	0.0	0.0	0.0
Sw.	0.0	0.0	0.0	1.1	4.4	2.8	4.3
Switz.	0.0	0.0	0.0	0.0	0.0	0.0	0.0
Turk.	0.0	0.0	0.0	0.0	0.0	0.0	0.0
UK	0.0	0.0	0.0	4.4	0.0	4.2	0.1
Can.	0.0	0.0	0.0	0.0	0.0	0.0	0.0
USA	0.0	0.0	0.0	0.0	0.0	0.0	0.0
Jap.	0.0	0.0	0.0	0.0	0.0	0.0	0.0
Austral.	4.0	3.9	3.4	2.8	5.8	4.9	5.5
NZ	0.0	0.0	0.0	1.3	0.0	0.0	0.7

Note: Taxes on payroll and work-force cover taxes paid by employers, employees, or the self-employed either as a proportion of payroll or as a fixed amount per person, and which are not earmarked for social security expenditure. General government consists of all departments, offices, organizations, and other bodies which are agents or instruments of the central, state, or local public authorities.

Sources: OECD, *Revenue Statistics 1965–1983* (1984); OECD, *Revenue Statistics 1965–1987* (1988).

TABLE 5.21. General government: taxes on goods and services as a percentage of total tax revenue

	1955	1960	1965	1970	1975	1980	1985
Aus.	38.9	38.8	37.4	37.4	34.5	31.3	32.6
Belg.	38.9	38.8	37.2	35.1	26.4	26.1	24.5
Den.	37.6	39.4	40.6	38.8	33.6	37.4	34.2
Fin.	43.4	46.1	43.6	40.5	34.7	40.3	36.2
Fr.	—	—	38.4	38.1	32.9	30.1	29.7
FRG	35.9	32.8	33.0	31.8	26.7	26.9	25.7
Gr.	—	—	52.2	50.2	48.3	40.1	42.6
Ice.	—	—	—	—	—	—	—
Ire.	52.4	54.8	52.6	52.4	46.5	43.7	44.4
It.	45.1	42.3	39.5	38.7	29.4	27.4	25.4
Lux.	—	—	24.8	20.5	20.8	19.8	23.9
Neth.	32.9	29.2	28.5	27.8	23.9	24.8	25.6
Norw.	44.5	42.8	41.0	42.8	37.6	35.4	37.5
Port.	40.7	39.3	44.2	44.6	40.8	44.9	42.6
Sp.	—	—	40.8	35.9	24.2	20.7	26.4
Sw.	28.6	32.3	31.2	28.2	24.4	24.0	26.4
Switz.	34.7	34.3	30.5	26.9	19.9	20.4	18.9
Turk.	59.1	53.0	54.0	49.4	41.3	29.2	36.0
UK	35.8	34.5	33.0	28.8	25.4	29.1	31.5
Can.	44.4	37.7	41.2	31.6	32.1	32.6	31.8
USA	22.1	21.5	21.9	19.2	18.5	16.7	17.8
Jap.	34.0	31.5	26.3	22.4	17.3	16.3	14.0
Austral.	36.9	37.0	34.5	32.0	29.0	30.9	32.4
NZ	32.2	31.3	27.9	27.2	24.2	22.4	23.2

Note: Taxes on goods and services cover all taxes and duties levied on the production, extraction, sale, transfer, leasing, or delivery of goods, and the rendering of services, or in respect of the use of goods or permission to use goods or to perform activities. General government consists of all departments, offices, organizations, and other bodies which are agents or instruments of the central, state, or local public authorities.

Sources: OECD, *Revenue Statistics 1965–1983* (1984); OECD, *Revenue Statistics 1965–1987* (1988).

TABLE 5.22. General government: social security contributions as a percentage of GDP

	1955	1960	1965	1970	1975	1980	1985
Aus.	7.1	7.4	8.7	9.1	10.7	13.0	13.6
Belg.	6.2	7.2	9.8	10.7	13.3	13.6	15.3
Den.	1.1	1.3	1.6	1.6	0.6	0.8	1.6
Fin.	2.1	1.9	0.9	1.5	3.0	2.9	3.5
Fr.	—	—	11.9	12.7	15.3	18.3	18.9
FRG	7.6	8.6	8.5	10.0	12.3	13.0	14.0
Gr.	—	—	5.5	6.6	6.7	9.6	12.0
Ice.	—	—	—	—	—	—	—
Ire.	1.0	1.1	1.7	2.6	4.4	5.1	5.7
It.	9.8	11.3	9.3	9.1	13.3	12.2	12.4
Lux.	—	—	9.9	8.6	11.7	11.8	10.6
Neth.	4.2	7.6	10.3	13.2	16.7	17.4	19.3
Norw.	1.3	2.8	4.0	6.3	11.1	9.9	11.1
Port.	2.7	3.0	4.0	5.5	8.6	8.6	9.1
Sp.	—	—	4.2	6.4	9.3	11.7	11.9
Sw.	0.5	1.2	4.3	6.0	8.6	14.1	13.4
Switz.	4.1	4.6	4.7	5.6	8.6	9.5	10.3
Turk.	0.6	0.8	0.9	1.1	2.0	1.1	2.9
UK	3.1	3.6	4.7	5.2	6.2	6.1	7.0
Can.	0.9	1.4	1.5	3.0	3.3	3.4	4.5
USA	2.6	3.8	4.3	5.6	7.3	7.9	8.6
Jap.	2.2	2.5	4.0	4.4	6.1	7.5	8.6
Austral.	0.0	0.0	0.0	0.0	0.0	0.0	0.0
NZ	6.1	5.6	0.0	0.0	0.0	0.0	0.0

Note: Social security contributions include all compulsory contributions as payments to institutions of general government providing for social security benefits. General government consists of all departments, offices, organizations, and other bodies which are agents or instruments of the central, state, or local public authorities.

Sources: OECD, *Revenue Statistics 1965–1983* (1984); OECD, *Revenue Statistics 1965–1987* (1988).

TABLE 5.23. General government: social security expenditure as a percentage of GDP

	1960	1965	1970	1975	1980	1983
Aus.	15.4	17.7	18.6	20.2	22.4	24.2
Belg.	15.3	16.1	18.1	23.6	25.9	28.0
Den.	11.1	12.2	16.4	22.4	26.9	27.9
Fin.	8.8	10.6	12.8	16.1	18.6	20.6
Fr.	13.2	15.6	15.3	24.1	26.8	29.4
FRG	15.4	16.6	17.0	23.5	23.8	24.3
Gr.	—	9.2	10.8	10.8	12.2	17.6
Ice.	—	—	—	—	—	—
Ire.	9.3	10.3	11.6	19.7	21.7	23.5
It.	11.7	14.8	16.3	23.1	18.2	25.7
Lux.	13.4	15.1	15.3	21.6	24.1	25.4
Neth.	11.1	15.5	20.0	26.8	28.6	31.9
Norw.	9.4	10.9	15.5	18.5	20.3	21.9
Port.	5.3	5.3	5.7	11.0	10.1	10.1
Sp.	—	—	—	11.7	16.1	17.7
Sw.	10.9	13.6	18.8	26.2	32.0	33.3
Switz.	7.5	8.5	10.1	15.1	13.8	14.6
Turk.	1.4	1.7	3.2	3.5	4.3	3.8
UK	10.8	11.7	13.8	16.2	17.7	20.5
Can.	9.2	9.4	11.8	14.7	15.1	16.5
USA	6.8	7.1	9.6	13.2	12.7	13.8
Jap.	4.9	5.1	5.4	7.6	10.9	12.0
Austral.	7.7	8.3	8.0	10.7	12.1	12.4
NZ	13.1	11.5	11.5	12.5	14.4	17.6

Note: Total social security expenditure includes expenditure on medical care, benefits in kind other than for medical care, and cash benefits. General government consists of all departments, offices, organizations, and other bodies which are agents or instruments of the central, state, or local public authorities.

Sources: 1960–1980: ILO (1985); 1983: ILO (1988).

TABLE 5.24. General government: health expenditure as a percentage of total government expenditure

	1960	1965	1970	1975	1980
Aus.	—	8.3	8.8	9.2	9.6
Belg.	8.7	9.1	9.6	9.8	9.7
Den.	—	—	12.7	12.5	10.4
Fin.	8.6	10.1	13.2	13.5	13.2
Fr.	7.4	9.5	11.3	12.8	13.4
FRG	10.0	9.7	11.4	13.9	14.0
Gr.	7.3	8.6	7.6	7.8	10.5
Ice.	—	—	—	—	—
Ire.	9.4	9.3	11.4	13.9	17.1
It.	10.7	12.1	14.3	13.9	14.2
Lux.	—	—	—	—	—
Neth.	4.4	8.6	12.5	11.7	11.4
Norw.	10.5	9.3	11.2	13.9	13.4
Port.	—	—	—	—	—
Sp.	—	—	—	—	—
Sw.	10.9	12.4	14.3	14.7	14.4
Switz.	—	—	15.2	16.6	—
Turk.	—	—	—	—	—
UK	10.2	10.2	10.4	11.0	11.7
Can.	8.1	10.5	14.2	14.0	13.7
USA	4.7	5.5	8.5	10.1	11.4
Jap.	7.8	14.5	15.2	15.1	15.4
Austral.	10.8	11.0	12.2	17.1	15.4
NZ	11.0	11.8	12.7	14.2	12.7

Note: Health expenditure includes expenditure on hospitals, clinics and medical, dental and paramedical practitioners, public health, medicaments, etc. By general government is meant central government, state or provincial government, local government, and social security funds. Total government expenditure includes current and capital outlays.

Source: OECD (1985c): annex C.

TABLE 5.25. General government: health expenditure as a percentage of GDP

	1960	1965	1970	1975	1980
Aus.	2.9	3.0	3.4	4.1	4.5
Belg.	2.1	2.9	3.5	4.5	5.5
Den.	3.2	4.2	5.2	5.9	5.8
Fin.	2.3	3.2	4.1	4.9	5.0
Fr.	2.5	3.6	4.3	5.5	6.1
FRG	3.2	3.6	4.2	6.6	6.5
Gr.	1.7	2.2	2.2	2.5	3.5
Ice.	2.4	2.8	4.1	6.7	6.7
Ire.	3.0	3.3	4.3	6.3	8.1
It.	3.2	4.1	4.8	5.8	6.0
Lux.	—	—	—	5.6	6.6
Neth.	1.3	3.0	5.1	5.9	6.5
Norw.	2.6	3.2	4.6	6.4	6.7
Port.	0.9	1.2	1.9	3.8	4.2
Sp.	—	1.4	2.3	3.6	4.3
Sw.	3.4	4.5	6.2	7.2	8.8
Switz.	—	2.3	—	4.7	4.7
Turk.	—	1.2	1.3	—	1.1
UK	3.4	3.6	3.9	5.0	5.2
Can.	2.4	3.1	5.1	5.7	5.4
USA	1.3	1.6	2.8	3.7	4.1
Jap.	1.8	2.7	3.0	4.0	4.6
Austral.	2.4	2.8	3.2	5.6	4.7
NZ	3.3	3.4	3.5	4.3	4.7

Note: Public expenditure on health care equals current general government expenditure on health care. By general government is meant central government, state or provincial government, local government, and social security funds.

Source: OECD (1985c): annex C.

TABLE 5.26. General government: total health expenditure as a percentage of GDP

	1960	1965	1970	1975	1980
Aus.	4.4	4.7	5.3	6.4	7.0
Belg.	3.4	3.9	4.1	5.5	6.3
Den.	3.6	4.8	6.1	6.5	6.8
Fin.	4.2	4.9	5.6	5.8	6.3
Fr.	4.3	5.3	6.1	7.6	8.5
FRG	4.8	5.1	5.6	8.1	8.1
Gr.	2.9	3.1	3.9	4.0	4.2
Ice.	5.9	6.1	8.7	—	7.7
Ire.	4.0	4.4	5.6	7.7	8.7
It.	3.9	4.6	5.5	6.7	6.8
Lux.	—	—	4.9	5.9	6.6
Neth.	3.9	4.4	6.0	7.7	8.3
Norw.	3.3	3.9	5.0	6.7	6.8
Port.	—	—	—	6.4	6.1
Sp.	—	2.7	4.1	5.1	5.9
Sw.	4.7	5.6	7.2	8.0	9.5
Switz.	3.3	3.8	5.2	7.1	7.2
Turk.	—	—	—	—	—
UK	3.9	4.2	4.5	5.5	5.8
Can.	5.5	6.1	7.2	7.4	7.3
USA	5.3	6.1	7.6	8.6	9.5
Jap.	3.0	4.5	4.6	5.7	6.4
Austral.	5.1	5.3	5.7	7.6	7.4
NZ	4.4	—	4.5	5.2	5.7

Note: Total health expenditure equals current general government expenditure on health care and household final consumption on medical care and health expenses. By general government is meant central government, state or provincial government, local government, and social security funds.

Source: OECD (1985*d*): table 2.

TABLE 5.27. Central government: final consumption as a percentage of general government final consumption

	1950	1955	1960	1965	1970	1975	1980	1985
Aus.	52.6	46.4	49.0	47.2	46.2	37.3	35.4	37.3
Belg.	75.8	71.2	77.6	77.5	77.4	77.0	73.7	73.6
Den.	—	—	45.9	45.0	—	—	31.3	31.1
Fin.	—	—	43.8	44.2	40.3	35.8	33.0	30.5
Fr.	—	—	77.4	75.5	75.6	73.6	72.0	69.6
FRG	30.1	32.9	30.2	32.8	24.9	21.0	19.2	18.9
Gr.	83.4	84.7	79.0	76.2	77.7	82.0	74.3	71.0
Ice.	88.5	67.0	66.5	62.3	64.2	68.0	54.6	55.8
Ire.	56.7	55.4	57.6	58.3	50.8	49.1	46.2	49.3
It.	72.8	70.1	68.0	68.9	57.6	54.7	54.9	53.0
Lux.	73.0	71.6	70.7	69.1	68.7	67.5	66.0	65.8
Neth.	—	55.2	46.7	42.2	47.5	45.1	45.0	46.4
Norw.	56.1	56.2	51.4	49.2	46.4	42.4	39.9	37.4
Port.	86.8	88.6	87.3	89.8	88.0	79.0	86.7	87.6
Sp.	—	—	—	77.4	77.6	70.8	67.7	50.5
Sw.	58.9	52.9	45.4	49.1	37.4	33.9	29.9	26.1
Switz.	—	—	27.3	27.7	27.8	23.1	23.4	24.6
Turk.	—	—	—	85.5	82.5	—	—	—
UK	70.7	71.8	67.2	63.0	59.3	56.4	59.1	61.0
Can.	50.1	60.7	46.2	38.7	27.3	24.9	23.5	25.0
USA	52.4	63.7	58.0	53.6	51.4	42.9	42.3	47.4
Jap.	—	—	42.7	39.7	27.5	23.7	24.1	24.4
Austral.	—	—	45.5	50.3	46.7	35.7	33.8	32.9
NZ	—	—	—	—	—	—	—	—

Note: Central government consists of all departments, offices, organizations, and other bodies classified under general government which are agencies or instruments of the central authority of a country, except separately organized social security funds.

Sources: 1950–65: OECD, *National Accounts 1950–1968* (1968); OECD, *National Accounts 1960–1977* (1979); 1970–80: OECD, *National Accounts 1964–1981* (1983); 1985: OECD, *National Accounts 1973–1985* (1987); OECD, *National Accounts 1975–1987* (1989).

TABLE 5.28. Central government: current disbursements as a percentage of general government current disbursements

	1950	1955	1960	1965	1970	1975	1980	1985
Aus.	58.2	55.0	55.2	55.2	56.4	55.8	51.6	54.9
Belg.	73.6	70.9	72.0	65.3	64.3	64.4	65.9	67.1
Den.	—	—	66.4	68.6	—	—	74.2	74.6
Fin.	—	—	66.6	67.1	60.1	60.7	62.2	59.5
Fr.	—	—	60.2	55.3	54.9	50.6	48.3	47.8
FRG	36.1	43.5	38.4	39.6	35.1	33.5	33.0	32.5
Gr.	82.1	71.8	66.2	65.6	63.8	72.8	66.6	60.9
Ice.	67.1	70.5	75.8	69.8	77.5	87.9	82.2	81.0
Ire.	82.0	80.4	82.6	83.3	81.8	81.9	85.2	86.2
It.	64.8	62.1	56.2	56.1	55.2	55.9	76.1	81.7
Lux.	69.8	62.5	60.0	56.6	57.3	56.1	55.4	58.8
Neth.	—	78.7	65.9	59.7	58.6	58.8	61.0	57.3
Norw.	75.4	75.0	76.9	77.6	83.5	82.6	93.0	87.3
Port.	78.1	77.8	77.4	77.9	79.3	69.6	72.8	77.4
Sp.	—	—	—	60.0	60.6	51.1	51.9	53.0
Sw.	76.1	70.6	62.2	64.7	61.6	60.8	57.7	58.9
Switz.	—	—	28.0	27.0	32.3	26.9	29.8	29.4
Turk.	—	—	—	90.1	80.7	—	—	—
UK	84.8	81.3	78.2	72.4	72.6	76.1	75.5	74.6
Can.	61.9	68.2	60.1	54.2	52.8	55.5	52.5	53.5
USA	61.4	65.5	60.1	57.9	54.5	48.9	48.8	52.3
Jap.	—	—	43.6	39.9	62.0	56.3	59.4	55.6
Austral.	—	—	80.9	79.5	79.3	82.5	80.3	77.6
NZ	—	—	—	—	—	—	—	—

Note: Central government consists of all departments, offices, organizations, and other bodies classified under general government which are agencies or instruments of the central authority of a country, except separately organized social security funds.

Sources: 1950–65: OECD, *National Accounts 1950–1968* (1968); OECD, *National Accounts 1960–1977* (1979); 1970–80: OECD, *National Accounts 1964–1981* (1983); 1985: OECD, *National Accounts 1973–1985* (1987); OECD, *National Accounts 1975–1987* (1989).

TABLE 5.29. Central government: current disbursements minus transfers to other levels of general government as a percentage of general government current disbursements

	1950	1955	1960	1965	1970	1975	1980	1985
Aus.	55.6	50.6	49.5	46.3	46.1	40.8	40.4	41.7
Belg.	59.0	56.5	56.3	51.6	50.4	47.3	47.5	50.3
Den.	—	—	38.9	41.1	—	—	32.9	41.0
Fin.	—	—	54.0	53.6	44.9	44.0	43.4	39.2
Fr.	—	—	56.0	50.6	48.1	43.7	40.6	40.5
FRG	30.5	33.0	24.4	26.8	26.2	24.0	23.7	24.4
Gr.	78.8	67.9	61.8	57.8	54.8	64.1	57.9	52.1
Ice.	60.2	62.5	61.9	55.0	46.9	56.6	54.2	55.1
Ire.	67.9	65.1	66.3	65.2	63.3	60.2	59.4	62.1
It.	58.3	53.8	48.4	43.8	44.3	40.6	44.6	46.8
Lux.	54.9	51.5	47.3	44.6	44.2	45.0	44.3	46.4
Neth.	—	51.1	35.6	29.2	31.0	29.4	29.6	29.8
Norw.	65.6	63.7	69.6	69.5	74.4	74.0	68.2	65.0
Port.	77.5	77.4	76.8	77.8	78.9	65.7	68.6	71.0
Sp.	—	—	—	58.9	57.6	48.3	45.1	39.0
Sw.	69.0	62.5	43.4	43.7	47.4	43.9	44.4	45.6
Switz.	—	—	25.1	22.5	20.5	16.2	17.4	17.6
Turk.	—	—	—	80.9	76.6	—	—	—
UK	72.9	70.6	65.2	58.1	55.5	55.4	58.5	59.8
Can.	54.8	61.2	50.0	43.9	40.6	43.0	41.1	42.9
USA	57.3	62.2	55.0	51.5	44.7	37.0	37.3	44.3
Jap.	—	—	—	—	28.4	24.0	27.9	28.2
Austral.	—	—	59.2	59.3	56.6	54.9	54.3	56.6
NZ	—	—	—	—	—	—	—	—

Note: Central government consists of all departments, offices, organizations, and other bodies classified under general government which are agencies or instruments of the central authority of a country, except separately organized social security funds.

Sources: 1950–65: OECD, *National Accounts 1950–1968* (1968); OECD, *National Accounts 1960–1977* (1979); 1970–80: OECD, *National Accounts 1964–1981* (1983); 1985: OECD, *National Accounts 1973–1985* (1987); OECD, *National Accounts 1975–1987* (1989).

TABLE 5.30. General government: expenditure by function as a percentage of GDP 1975 and 1980

	Defence		Education		Health	
	1975	1980	1975	1980	1975	1980
Aus.	1.1	1.1	3.5	3.9	4.1	4.4
Belg.	2.7	2.8	—	—	—	—
Den.	2.5	2.5	—	6.0	5.5	5.5
Fin.	1.4	1.4	4.8	4.9	3.7	3.9
Fr.	3.2	3.3	4.9	5.1	0.4	0.5
FRG	3.1	2.7	4.0	4.1	6.1	5.9
Gr.	6.8	5.8	1.9	2.2	1.1	1.7
Ice.	—	0.0	—	3.6	—	5.4
Ire.	—	—	—	—	—	—
It.	1.9	1.9	4.0	4.9	3.4	3.6
Lux.	—	—	—	—	—	—
Neth.	3.1	3.0	—	—	—	—
Norw.	3.2	2.8	5.4	5.0	3.6	4.0
Port.	4.8	2.9	5.5	3.4	4.2	2.6
Sp.	1.7	—	1.3	—	1.2	—
Sw.	3.3	3.1	5.2	—	5.8	7.3
Switz.	—	—	—	—	—	—
Turk.	3.5[a]	—	3.5[a]	—	1.4[a]	—
UK	4.9	4.9	5.2	4.3	4.7	4.8
Can.	—	—	—	—	—	—
USA	5.7	5.3	5.2	4.6	1.2	1.0
Jap.	0.9	0.9	3.9	3.7	0.4	0.4
Austral.	2.2	2.5	4.7	4.7	3.1	4.7
NZ	—	—	—	—	—	—

[a] Figure for 1972.

Note: General government consists of all departments, offices, organizations, and other bodies which are agents or instruments of the central, state, or local public authorities.

Sources: OECD, *National Accounts 1964–1981* (1983): table 3a; OECD, *National Accounts 1973–1985* (1987): table 5.

TABLE 5.31. General government:
total social expenditure as a percentage of total
government expenditure

	1960	1965	1970	1975	1980
Aus.	—	53.9	56.5	55.4	57.3
Belg.	57.4	65.3	68.2	76.2	72.8
Den.	—	—	63.3	68.1	61.9
Fin.	58.2	57.0	63.9	63.8	66.4
Fr.	—	—	—	56.0	61.4
FRG	65.3	63.4	63.2	68.8	66.1
Gr.	36.2	38.9	38.8	34.1	40.0
Ice.	—	—	—	—	—
Ire.	36.4	38.5	45.1	49.7	52.3
It.	56.4	60.0	64.2	62.5	63.7
Lux.	—	—	—	—	—
Neth.	54.2	65.3	71.7	73.6	62.5
Norw.	44.3	46.2	55.0	56.3	55.8
Port.	—	—	—	—	—
Sp.	—	—	—	—	—
Sw.	49.3	51.3	52.9	54.7	53.2
Switz.	—	—	59.3	65.8	—
Turk.	—	—	—	—	—
UK	42.0	46.0	48.9	49.2	49.6
Can.	41.5	46.5	52.3	53.6	52.6
USA	38.7	43.3	47.5	57.8	58.5
Jap.	47.1	50.7	47.9	53.9	56.2
Austral.	44.9	42.4	44.3	57.9	60.8
NZ	43.8	44.8	46.7	53.1	51.2

Notes: By general government is meant central government, state or provincial government, local government, and social security funds. Total government expenditure includes current and capital outlays. Total social expenditure includes expenditure on education, health, pensions, unemployment compensation, and other social expenditure such as sickness, maternity, or temporary disablement benefits.

Source: OECD (1985c): annex C.

TABLE 5.32. General government: total social expenditure as a percentage of GDP

	1960	1965	1970	1975	1980
Aus.	17.9	19.7	21.6	24.5	27.1
Belg.	17.6	21.4	25.3	34.5	38.2
Den.	—	—	26.2	32.4	33.3
Fin.	15.4	17.7	19.9	23.3	25.4
Fr.	—	—	—	24.2	28.3
FRG	20.5	22.4	23.5	32.6	30.8
Gr.	8.5	10.1	10.9	10.6	13.4
Ice.	—	—	—	—	—
Ire.	11.7	13.9	17.1	23.1	25.7
It.	16.8	20.1	21.4	26.0	26.9
Lux.	—	—	—	—	—
Neth.	16.2	23.0	29.1	37.1	35.6
Norw.	11.7	15.8	22.5	26.2	27.2
Port.	—	—	—	—	—
Sp.	—	—	—	—	—
Sw.	15.4	18.6	23.0	26.8	32.5
Switz.	8.0	10.3	12.6	19.1	20.0
Turk.	—	—	—	—	—
UK	13.8	16.2	18.5	22.4	22.0
Can.	12.1	13.6	18.7	21.8	21.0
USA	10.9	12.3	15.7	20.8	20.7
Jap.	8.0	9.4	9.3	14.2	16.9
Austral.	10.2	11.0	11.5	18.8	18.6
NZ	13.0	12.7	12.7	16.3	19.4

Notes: By general government is meant central government, state or provincial government, local government, and social security funds. Total social expenditure includes expenditure on education, health, pensions, unemployment compensation, and other social expenditure as sickness, maternity, or temporary disablement benefits.

Source: OECD (1985c): annex C.

TABLE 5.33. General government:
educational expenditure as a percentage of total
government expenditure

	1960	1965	1970	1975	1980
Aus.	—	6.1	7.0	7.9	8.0
Belg.	14.6	16.2	16.2	17.2	15.2
Den.	—	—	17.1	17.1	14.3
Fin.	24.9	21.2	20.3	18.1	16.1
Fr.	—	—	—	13.5	12.4
FRG	7.6	9.8	10.8	11.4	11.0
Gr.	6.9	7.3	6.7	6.2	7.1
Ice.	—	—	—	—	—
Ire.	9.3	11.9	13.7	13.0	13.2
It.	12.5	11.9	13.4	11.9	13.3
Lux.	—	—	—	—	—
Neth.	15.1	16.7	16.4	15.0	12.7
Norw.	14.6	16.6	15.6	14.5	12.8
Port.	—	—	—	—	—
Sp.	—	—	—	—	—
Sw.	14.7	14.9	14.3	11.5	10.7
Switz.	—	—	19.5	19.1	—
Turk.	—	—	—	—	—
UK	11.0	12.8	14.0	14.9.	12.7
Can.	10.3	15.9	19.3	15.7	15.1
USA	12.8	15.7	16.0	17.6	16.1
Jap.	23.3	21.1	18.5	18.6	16.5
Austral.	12.3	13.3	16.0	18.9	19.0
NZ	9.0	11.1	12.8	14.7	11.1

Notes: By general government is meant central government, state or provincial government, local government, and social security funds. Total government expenditure includes current and capital outlays. Education includes expenditure on pre-primary, primary, secondary, tertiary, education affairs and services, and subsidiary services to education.

Source: OECD (1985c): annex C.

TABLE 5.34. General government: educational expenditure as a percentage of GDP

	1960	1965	1970	1975	1980
Aus.	2.0	2.2	2.7	3.5	3.8
Belg.	4.5	5.3	6.0	7.8	8.0
Den.	—	—	7.1	8.1	7.7
Fin.	6.6	6.6	6.3	6.6	6.2
Fr.	—	—	—	5.8	5.7
FRG	2.4	3.5	4.0	5.4	5.1
Gr.	1.6	1.9	1.9	1.9	2.4
Ice.	—	—	—	—	—
Ire.	3.0	4.3	5.2	6.1	6.5
It.	3.7	4.0	4.5	5.0	5.6
Lux.	—	—	—	—	—
Neth.	4.5	5.9	6.7	7.6	7.2
Norw.	3.9	5.7	6.4	6.7	6.3
Port.	—	—	—	—	—
Sp.	—	—	—	—	—
Sw.	4.6	5.4	6.2	5.6	6.5
Switz.	3.1	3.5	4.1	5.6	5.5
Turk.	—	—	—	—	—
UK	3.6	4.5	5.3	6.8	5.6
Can.	3.0	4.7	6.9	6.4	6.0
USA	3.6	4.5	5.3	6.3	5.7
Jap.	4.0	3.9	3.6	4.9	5.0
Austral.	2.8	3.5	4.2	6.2	5.8
NZ	2.7	3.1	3.5	4.5	4.2

Notes: By general government is meant central government, state or provincial government, local government, and social security funds. Education includes expenditure on pre-primary, primary, secondary, tertiary, education affairs and services, and subsidiary services to education.

Source: OECD (1985c): annex C.

TABLE 5.35. Central government: expenditure by function as a percentage of total expenditure, 1975

	Defence	Education	Health	Social security	Housing	Economic	Other services
Aus.	3.3	10.6	12.3	45.5	3.3	12.7	12.4
Belg.	6.3	16.2	1.6	43.7	1.6	15.1	15.4
Den.	7.0	11.5	3.5	43.5	2.0	10.5	22.0
Fin.	5.1	14.6	10.6	26.8	1.4	30.4	11.0
Fr.	7.6	9.9	15.0	40.8	3.2	9.5	14.0
FRG	10.5	1.1	19.6	48.9	0.3	8.7	10.9
Gr.	20.1	8.5	8.0	26.9	2.0	20.2	14.3
Ice.	—	11.9	17.9	13.1	4.1	33.5	19.5
Ire.	—	—	—	—	—	—	—
It.	4.9	12.3	16.4	32.8	2.0	18.4	13.2
Lux.	2.2	8.6	2.3	46.7	2.6	19.0	18.6
Neth.	6.4	14.9	11.9	36.4	2.1	9.3	19.0
Norw.	9.0	9.3	12.9	31.5	5.6	20.6	11.1
Port.	14.4	11.6	4.8	29.0	4.0	11.8	24.4
Sp.	5.8	7.8	0.9	51.6	1.7	14.0	18.1
Sw.	10.2	11.9	3.0	44.2	1.3	10.2	19.1
Switz.	10.8	4.1	10.4	48.1	1.3	14.8	10.7
Turk.	15.9	23.0	3.0	1.8	1.8	42.3	12.2
UK	13.7	2.6	12.9	21.7	3.5	12.4	33.3
Can.	7.3	3.2	7.7	35.0	2.0	18.7	26.1
USA	24.6	3.4	9.3	36.6	2.9	8.7	14.4
Jap.	—	—	—	—	—	—	—
Austral.	10.1	10.8	8.1	24.5	1.9	9.0	35.6
NZ	4.9	16.2	15.4	24.1	0.8	20.3	18.3

Notes: The central government covers all government departments, offices, establishments, and other bodies that are agencies or instruments of the central authority of a country and includes decentralized agencies, departmental enterprises, and relevant non-profit institutions attached to the central authority. Also included are social security funds, if operating nationally. Total expenditure by function comprises current expenditure and capital expenditure. Defence comprises expenditure intended mainly for military purposes. Education comprises provision, management, inspection, and support of pre-primary, primary, and secondary schools, universities and colleges, and technical, vocational, and other training institutions. Health comprises expenditures on hospitals, medical and dental centres, and clinics with a major medical component; provision of national health and medical insurance schemes. Social security and welfare comprise expenditures to compensate for temporary loss of income of the sick and temporarily disabled; to cover payments to the elderly, the permanently disabled, and the unemployed. Housing and community amenities comprise expenditure on housing, such as income-related schemes. Economic services comprise expenditure for: agriculture; industry; electricity, gas, and water; transport and communications; other economic services such as tourism, etc. 'Other' comprises expenditure on the general administration of government, etc.

Source: World Bank (1984): vol. i, economic data sheet 2.

TABLE 5.36. *General government: military expenditure as a percentage of GDP*

	1950	1955	1960	1965	1970	1975	1980	1985
Aus.	0.7	0.2	1.2	1.2	1.1	1.2	1.2	1.3
Belg.	—	3.8	3.6	3.2	2.9	3.1	3.3	3.0
Den.	1.7	3.2	2.7	2.8	2.4	2.4	2.4	2.2
Fin.	1.8	1.6	1.7	1.7	1.4	1.4	1.9	2.0
Fr.	5.5	6.4	6.5	5.2	4.2	3.8	4.0	4.0
FRG	4.4	4.1	4.0	4.3	3.3	3.6	3.3	3.2
Gr.	6.0	5.1	4.9	3.5	4.8	6.5	5.7	7.0
Ice.	—	—	—	—	—	—	—	—
Ire.	1.3	1.6	1.6	1.4	1.3	1.8	1.9	1.6
It.	4.3	3.7	3.3	3.3	2.7	2.5	2.1	2.3
Lux.	1.3	3.2	1.0	1.4	0.8	1.0	1.0	1.1
Neth.	4.8	5.7	4.1	4.0	3.5	3.4	3.1	3.1
Norw.	2.4	3.9	2.9	3.4	3.5	3.2	2.9	3.1
Port.	3.8	4.2	4.2	6.2	7.1	5.3	3.5	3.2
Sp.	—	2.2	2.2	1.8	1.6	1.7	3.1	3.3
Sw.	3.5	4.5	4.0	4.1	3.6	3.4	2.9	2.5
Switz.	2.6	2.8	2.5	2.7	2.2	2.0	1.9	1.9
Turk.	6.2	5.6	5.1	5.0	4.3	4.6	4.9	4.9
UK	6.6	8.2	6.5	5.9	4.8	5.0	4.9	5.3
Can.	2.6	6.3	4.2	2.9	2.4	1.9	1.8	2.1
USA	5.1	10.2	9.0	7.6	8.0	6.0	5.4	6.6
Jap.	—	1.8	1.1	0.9	0.8	0.9	0.9	1.0
Austral.	3.0	3.8	2.7	3.4	3.5	2.8	2.6	2.8
NZ	1.5	2.5	2.1	2.1	1.9	1.7	1.9	2.0

Note: General government consists of all departments, offices, organizations, and other bodies which are agents or instruments of the central, state, or local public authorities.

Sources: SIPRI (1980): table 1A.4; SIPRI (1988): table 6A.3.

TABLE 5.37. General government: official development assistance as a percentage of donor GNP

	1960	1965	1970	1975	1980	1985
Aus.	—	0.11	0.14	0.11	0.24	0.37
Belg.	0.88	0.63	0.47	0.68	0.49	0.52
Den.	0.09	0.13	0.47	0.76	0.82	0.91
Fin.	—	0.02	0.08	0.23	0.24	0.44
Fr.	1.35	0.76	0.62	0.66	0.64	0.71
FRG	0.31	0.40	0.35	0.41	0.43	0.42
Gr.	—	—	—	—	—	—
Ice.	—	—	—	—	—	—
Ire.	—	0.00	0.00	0.11	0.19	0.22
It.	0.22	0.08	0.13	0.09	0.16	0.26
Lux.	—	—	—	—	—	—
Neth.	0.31	0.36	0.62	0.84	0.99	0.85
Norw.	0.11	0.16	0.47	0.89	1.12	1.12
Port.	—	—	—	—	—	—
Sp.	—	—	—	—	—	—
Sw.	0.05	0.18	0.43	0.98	0.92	0.87
Switz.	0.04	0.09	0.15	0.23	0.24	0.29
Turk.	—	—	—	—	—	—
UK	0.56	0.48	0.40	0.42	0.39	0.31
Can.	0.19	0.19	0.44	0.58	0.41	0.46
USA	0.53	0.57	0.31	0.26	0.26	0.23
Jap.	0.24	0.28	0.23	0.24	0.31	0.29
Austral.	0.37	0.50	0.56	0.60	0.47	0.46
NZ	—	0.00	0.06	0.51	0.35	0.23

Note: General government consists of of all departments, offices, organizations, and other bodies which are agents or instruments of the central, state, or local public authorities.

Sources: World Bank, *World Development Report* (1984): table 18; World Bank, *World Development Report* (1988): table 21.

TABLE 5.38. General government: total tax revenues as a percentage of GDP

	1955	1960	1965	1970	1975	1980	1985
Aus.	30.0	30.6	34.7	35.7	38.7	41.3	42.9
Belg.	24.0	26.5	31.2	35.8	41.8	44.7	46.6
Den.	23.4	25.4	29.9	40.4	41.4	45.5	48.7
Fin.	26.8	27.7	30.1	32.2	36.2	35.3	36.8
Fr.	—	—	35.0	35.6	37.4	42.5	44.5
FRG	30.8	31.3	31.6	32.9	36.0	37.8	37.9
Gr.	—	—	20.6	24.3	24.6	28.6	35.2
Ice.	—	—	—	—	—	—	—
Ire.	22.5	21.9	26.0	31.2	32.1	35.9	39.0
It.	30.5	34.4	27.3	27.9	29.0	33.2	34.7
Lux.	—	—	30.5	30.3	38.5	40.4	42.9
Neth.	26.3	30.1	33.7	37.9	43.6	45.7	45.1
Norw.	28.3	31.2	33.2	39.2	44.8	47.1	47.4
Port.	15.4	16.3	18.5	23.1	24.8	29.3	31.5
Sp.	—	—	14.7	17.2	19.6	24.1	28.8
Sw.	25.5	27.2	35.8	40.2	43.9	49.4	50.6
Switz.	19.2	21.3	20.7	23.8	29.6	30.8	32.0
Turk.	11.9	11.5	15.0	17.7	20.7	19.0	19.7
UK	29.8	28.5	30.6	37.3	35.7	36.0	38.1
Can.	21.7	24.2	25.9	32.0	32.9	32.7	32.9
USA	23.6	26.5	26.3	29.8	29.6	30.4	29.2
Jap.	17.1	18.2	18.4	19.7	21.0	25.9	28.0
Austral.	22.5	23.4	24.3	25.3	29.0	30.3	30.8
NZ	27.0	27.4	25.0	26.9	29.6	31.0	33.8

Note: Total tax revenues include taxes on income, profits, and capital gains; social security contributions; taxes on payroll and work-force; taxes on property; taxes on goods and services; and other taxes. General government consists of all departments, offices, organizations, and other bodies which are agents or instruments of the central, state, or local public authorities.

Sources: OECD, *Revenue Statistics 1965–1983* (1984); OECD, *Revenue Statistics 1965–1987* (1988).

TABLE 5.39. General government: taxes on income and profits as a percentage of total tax revenue

	1955	1960	1965	1970	1975	1980	1985
Aus.	23.8	23.5	25.7	25.2	26.2	26.7	26.4
Belg.	31.1	29.9	27.6	31.4	39.3	41.0	40.6
Den.	47.5	44.8	45.9	51.2	59.0	55.0	56.8
Fin.	42.8	41.1	44.0	47.6	52.3	49.1	51.0
Fr.	—	—	15.9	18.3	17.6	18.0	17.6
FRG	22.0	31.9	33.8	32.3	34.5	35.3	34.8
Gr.	—	—	9.8	12.3	13.7	19.3	17.4
Ice.	—	—	—	—	—	—	—
Ire.	23.8	21.3	25.7	27.1	30.0	36.5	34.5
It.	12.7	15.7	17.8	17.4	21.5	32.1	36.8
Lux.	—	—	35.8	43.3	43.1	44.6	44.5
Neth.	39.6	39.4	35.6	33.4	34.8	32.9	26.4
Norw.	46.5	44.2	43.5	38.4	34.4	41.3	39.3
Port.	27.7	27.8	24.6	23.7	17.4	19.7	25.9
Sp.	—	—	24.5	20.2	22.0	26.0	28.0
Sw.	66.9	61.3	54.9	54.2	50.5	43.5	42.0
Switz.	34.9	35.5	38.2	40.8	43.8	41.4	40.9
Turk.	34.9	39.3	29.6	33.4	42.3	59.1	37.0
UK	40.6	37.5	36.8	40.4	44.4	37.7	38.7
Can.	39.2	39.6	39.3	44.6	47.5	46.6	44.1
USA	53.4	49.9	46.3	47.9	43.8	47.1	42.8
Jap.	43.0	45.1	43.9	47.7	44.6	46.1	45.8
Austral.	49.0	47.3	50.4	54.3	55.9	55.9	54.3
NZ	33.9	35.7	60.5	61.1	66.5	70.2	69.0

Note: Taxes on income, profits, and capital gains cover taxes levied on the net income or profits of individuals and enterprises. Also covered are taxes levied on the capital gains of individuals and enterprises, and gains from gambling. General government consists of all departments, offices, organizations, and other bodies which are agents or instruments of the central, state, or local public authorities.

Sources: OECD, *Revenue Statistics 1965–1983* (1984); OECD, *Revenue Statistics 1965–1987* (1988).

TABLE 5.40. General government: deficit as a percentage of GDP

	1950	1955	1960	1965	1970	1975	1980	1985
Aus.	6.7	6.5	6.0	7.3	6.6	4.3	3.3	2.5
Belg.	-1.3	0.1	-1.2	0.8	2.2	-0.7	-4.9	-5.8
Den.	3.7	4.3	5.8	5.5	7.1	2.6	-0.0	0.3
Fin.	10.6	9.5	9.7	7.8	6.0	6.6	3.5	2.8
Fr.	5.9	3.2	3.8	4.8	4.3	1.1	2.5	-0.9
FRG	3.3	7.8	7.7	5.8	5.8	-0.7	1.8	2.0
Gr.	-4.1	1.9	2.5	2.1	4.4	0.7	0.1	-10.5
Ice.	7.7	6.9	13.1	8.5	10.1	7.2	7.1	5.1
Ire.	0.5	0.4	0.1	0.4	1.1	-6.7	-6.7	-6.1
It.	0.3	1.6	3.2	0.7	0.2	-7.1	-4.0	-6.6
Lux.	9.2	3.1	7.0	5.6	6.4	7.6	5.7	5.2
Neth.	9.1	3.4	5.4	3.8	4.3	2.1	0.8	-0.8
Norw.	7.7	6.4	6.5	5.8	7.0	7.8	8.9	12.1
Port.	3.7	3.2	2.3	2.6	4.8	-2.5	-2.3	-3.5
Sp.	—	—	4.4	3.5	3.8	3.1	0.6	-1.5
Sw.	2.7	6.3	6.3	10.2	9.9	5.8	-0.4	-1.4
Switz.	6.1	5.5	6.4	4.1	5.2	3.3	3.6	3.5
Turk.	—	—	—	4.4	7.2	—	—	—
UK	3.4	1.6	0.3	2.3	7.6	-0.3	-1.4	-1.2
Can.	4.9	2.6	1.5	4.3	3.0	0.1	-0.5	-5.0
USA	4.0	2.5	2.5	2.1	0.0	-3.1	-0.7	-4.2
Jap.	7.3	4.4	7.1	6.1	6.7	3.2	2.6	4.3
Austral.	—	—	6.5	6.7	5.9	3.4	3.0	-1.8
NZ	—	—	—	—	—	—	—	—

Note: The deficit is current receipts minus current disbursements. General government consists of all departments, offices, organizations, and other bodies which are agents or instruments of the central, state, or local public authorities.

Sources: 1950–65: OECD, *National Accounts 1950–1968* (1970); OECD, *National Accounts 1960–1977* (1979); 1970–80: OECD, *National Accounts 1964–1981* (1981); 1985: OECD, *National Accounts 1973–1985* (1987); OECD, *National Accounts 1975–1987* (1989).

TABLE 5.41. Central government: deficit as a percentage of GDP

	1950	1955	1960	1965	1970	1975	1980	1985
Aus.	0.9	4.0	3.2	3.2	2.3	−0.5	−0.0	−0.7
Belg.	−1.3	−0.2	−2.0	−0.3	0.8	−1.6	−3.9	−6.7
Den.	—	—	4.7	4.3	6.0	0.2	−2.7	−1.7
Fin.	—	—	6.8	4.8	5.1	3.3	1.5	1.7
Fr.	—	—	2.8	3.7	2.9	−0.2	0.9	−2.2
FRG	0.8	3.9	2.9	2.0	2.9	−0.6	0.4	0.6
Gr.	−5.6	0.7	1.3	0.3	1.8	−2.1	−2.1	−8.3
Ice.	6.4	4.7	10.3	5.2	7.2	2.8	4.8	3.2
Ire.	0.3	0.5	−0.1	0.5	1.5	−5.6	−5.6	−5.5
It.	−0.2	0.6	2.7	0.5	0.4	−2.6	−4.7	−7.5
Lux.	5.8	1.7	4.0	2.8	4.0	5.6	4.0	4.0
Neth.	—	2.3	4.1	3.4	3.5	2.0	0.7	−1.6
Norw.	5.8	4.3	4.5	3.8	4.5	5.9	7.6	−2.3
Port.	1.8	1.2	1.2	1.2	2.1	−2.5	−2.6	−4.8
Sp.	—	—	—	3.1	2.6	2.3	0.3	−2.2
Sw.	2.1	3.3	5.1	4.2	3.9	1.5	−5.3	−4.5
Switz.	—	—	3.3	2.6	1.7	0.7	0.5	0.5
Turk.	—	—	—	1.4	4.8	—	—	—
UK	2.3	1.6	0.1	2.1	7.4	−1.3	−1.3	−1.2
Can.	3.2	0.6	0.0	1.5	0.6	−1.7	−3.1	−6.0
USA	3.2	1.0	0.6	−0.1	−2.0	−3.7	−1.8	−4.9
Jap.	—	—	8.3	5.7	2.3	−1.2	−2.0	−1.5
Austral.	—	—	5.0	4.1	4.6	1.2	1.6	−0.5
NZ	—	—	—	—	—	—	—	—

Note: The deficit is current receipts minus current disbursements. Central government consists of all departments, offices, organizations, and other bodies classified under general government which are agencies or instruments of the central authority of a country, except separately organized social security funds.

Sources: 1950–65: OECD, *National Accounts 1950–1968* (1970), OECD, *National Accounts 1960–1977* (1979); 1970–80: OECD, *National Accounts 1964–1981* (1981); 1985: OECD, *National Accounts 1973–1985* (1987); OECD, *National Accounts 1975–1987* (1989).

Section 6: Government Structures

TABLES

6.1   Chamber systems: 1980s
6.2   Electoral systems in the 1980s: elections to the lower houses
6.3   Elections: male and female suffrage and current voting age
6.4   Constitutional development: dates of independence and current constitutions
6.5   Human rights, freedom, and democracy
6.6   Structure of government: levels and number of units in the 1980s
6.7   Elected governments in Western Europe, 1945–1989
6.8   Elected governments in Western Europe formed in the 1940s
6.9   Elected governments in Western Europe formed in the 1950s
6.10 Elected governments in Western Europe formed in the 1960s
6.11 Elected governments in Western Europe formed in the 1970s
6.12 Elected governments in Western Europe formed in the 1980s

TABLE 6.1. Chamber systems: 1980s

	Upper house	Seats	Period of office (years)	Lower house	Seats	Period of office (years)
Aus.	Bundesrat	63	5/6	Nationalrat	183	4
Belg.	Sénat	183	4	Chambre des représentants	212	4
Den.				Folketing	179	4
Fin.				Eduskunta	200	4
Fr.	Sénat	319	9	Assemblée nationale	577	5
FRG	Bundesrat	45	Varies	Bundestag	520	4
Gr.				Vouli	300	4
Ice.				Alþingi	60	4
Ire.	Seanad	60	5	Dáil	166	5
It.	Senato	323	5	Camera dei deputati	630	5
Lux.				Chambre des députés	60	5
Neth.	Eerste Kamer	75	4	Tweede Kamer	150	4
Norw.				Storting	165	4
Port.				Assembleia da Republica	250	4
Sp.	Senado	257	4	Congreso de diputados	350	4
Sw.				Riksdag	349	3
Switz.	Ständerat	46	4	Nationalrat	200	4
Turk.				Büyük Millet Meclisi	400	5
UK	House of Lords	1 180	Life	House of Commons	650	5 (max.)
Can.	Senate	104	Until retirement	House of Commons	195	5
USA	Senate	100	6	House of representatives	435	2
Jap.	House of councillors	252	6	House of representatives	512	4
Austral.	Senate	76	6	House of representatives	148	3
NZ				House of representatives	97	3

Sources: Mackie and Rose (1982); Interparliamentary Union (1986); Encyclopaedia Britannica (1988).

Government Structures

TABLE 6.2. Electoral systems in the 1980s: elections to the lower houses

	Electoral formula	Electoral district magnitude	Proportionality index
Aus.	PR: Hare (higher level seat allocation: d'Hondt)	20.3	98
Belg.	PR: Hare (higher level seat allocation: d'Hondt)	7.1	91
Den.	PR: Sainte Lague (higher level seat allocation: largest remainder)	9.7	99
Fin.	PR: d'Hondt	13.3	98
Fr.	Majority and plurality (PR in 1986)	1.0	79
FRG	Plurality (higher level seat allocation: PR: d'Hondt)	2.0	98
Gr.	PR: Hagenbach (higher level seat allocation: Hare quota)	5.4	88
Ice.	PR: d'Hondt	6.7	97
Ire.	PR: STV	4.0	96
It.	PR: imperiali quota (higher level seat allocation: largest remainder)	20.3	95
Lux.	PR: Hagenbach	14.8	90
Neth.	PR: d'Hondt	150.0	96
Norw.	PR: Sainte Lague	8.2	91
Port.	PR: d'Hondt	12.5	94
Sp.	PR: d'Hondt	6.7	83
Sw.	PR: Sainte Lague (higher level seat allocation: largest remainder)	12.0	98
Switz.	PR: Hagenbach	7.7	96
Turk.	PR	6.0	71
UK	Plurality	1.0	85
USA	Plurality	1.0	94
Can.	Plurality	1.0	88
Jap.	Plurality	3.9	91

	Electoral formula	Electoral district magnitude	Proportionality index
Austral.	Majority: alternative vote	1.0	87
NZ	Plurality	1.0	80

Note: Electoral district magnitude equals average number of seats per constituency in the whole country. The proportionality index is calculated by summing the difference between each party's share of seats and its share of votes, dividing by two, and subtracting the result from 100. Higher level seat allocation means allocation of seats at a higher level than the constituency, i.e. regional or national level.

Source: Mackie and Rose (1982): table A5; *European Journal of Political Research* (various years); Lijphart (1984).

TABLE 6.3. Elections: male and female suffrage and current voting age

	Completion of male suffrage	Completion of female suffrage	Current voting age
Aus.	1907	1918	19
Belg.	1893	1948	18
Den.	1901	1915	18
Fin.	1906	1906	18
Fr.	1848	1944	18
FRG	1869	1919	18
Gr.	1877	1952	20
Ice.	1915	1915	20
Ire.	1918	1918	18
It.	1912	1946	18
Lux.	1919	1919	18
Neth.	1917	1919	18
Norw.	1897	1913	18
Port.	1911	1974	18
Sp.	1869	1976	18
Sw.	1909	1921	18
Switz.	1919	1971	18
Turk.	1923	1934	20
UK	1918	1928	18
Can.	1917	1918	18
USA	1870	1920	18
Jap.	1925	1945	20
Austral.	1901	1902	18
NZ	1879	1893	18

Sources: Nohlen (1978): 37; Mackie and Rose (1982); Wallechinsky *et al.* (1980).

TABLE 6.4. Constitutional development: dates of independence and current constitutions

	Date of independence	Date of current constitution
Aus.	1918	1920
Belg.	1830	1831
Den.	c.800	1953
Fin.	1917	1919
Fr.	843	1958
FRG	1955	1949
Gr.	1830	1975
Ice.	1944	1944
Ire.	1921	1937
It.	1861	1948
Lux.	1867	1868
Neth.	1814	1814
Norw.	1905	1814
Port.	c.1140	1976
Sp.	1492	1978
Sw.	c.836	1975
Switz.	1499	1874
Turk.	1923	1982
UK	1066	[1688]
Can.	1867	1982
USA	1776	1789
Jap.	c.660 BC	1947
Austral.	1901	1900
NZ	1907	1852

Source: Encyclopaedia Britannica (1988).

TABLE 6.5. Human rights, freedom, and democracy

	Human rights (1)		Freedom (2)			Democracy Bollen (3)		Vanhanen (4)		
	1980	1985	1975	1980	1985	1960	1965	1950–9	1960–9	1970–9
Aus.	92	96	2	2	2	97.2	97.1	34.7	33.5	30.8
Belg.	92	96	2	2	2	99.7	99.9	32.0	34.2	38.0
Den.	96	98	2	2	2	99.9	99.9	29.7	33.6	40.3
Fin.	96	98	4	4	4	97.3	97.3	28.0	28.6	19.4
Fr.	88	94	3	3	3	89.7	90.8	28.8	17.1	33.4
FRG	91	97	2	3	3	88.0	88.6	31.6	30.7	32.8
Gr.	80	94	4	4	4	88.0	82.8	17.4	13.4	10.3
Ice.	—	—	2	2	2	100.0	100.0	—	—	—
Ire.	86	87	3	2	2	94.8	97.2	24.0	23.1	24.2
It.	88	87	3	4	2	97.0	96.8	32.9	36.5	38.4
Lux.	—	—	3	2	2	100.0	97.7	—	—	—
Neth.	94	98	2	2	2	99.9	99.7	35.9	37.1	39.8
Norw.	95	97	2	2	2	99.9	99.9	28.1	29.0	32.9
Port.	86	93	8	4	3	41.6	39.0	0.4	0.7	4.6
Sp.	78	86	10	4	3	10.7	10.4	0.0	0.0	3.0
Sw.	94	98	2	2	2	99.9	99.9	29.1	29.6	35.9
Switz.	92	95	2	2	2	99.6	99.7	14.7	12.9	22.3
Turk.	43	40	5	5	8	59.1	76.4	17.7	14.8	18.1
UK	95	94	2	2	2	99.3	99.1	28.0	26.9	30.0
Can.	94	96	2	2	2	99.9	99.5	20.9	22.9	24.4
USA	92	90	2	2	2	94.6	92.4	16.7	17.5	17.6

	Human rights (1)		Freedom (2)			Democracy					
						Bollen (3)			Vanhanen (4)		
	1980	1985	1975	1980	1985	1960	1965	1950–9	1960–9	1970–9
Jap.	92	88	3	3	2	99.3	99.8	22.8	20.2	26.5
Austral.	93	94	2	2	2	100.0	99.9	27.9	27.8	30.1
NZ	96	96	2	2	2	100.0	100.0	27.2	25.5	27.5

Notes: (1) This index represents a rating of human rights; the higher the index, the higher the standard of human rights. (2) This index of freedom summarizes ratings on political rights and civil liberties; the higher the index, the lower the level of freedom. (3) This index of democracy incorporates measures of press freedom, freedom of group opposition, government sanctions, fairness of elections, executive selection, and legislature selection; the higher the index, the higher the level of democracy. (4) This index of democracy incorporates measures of competition multiplied with measures of participation; the higher the index, the higher the level of democracy.

Sources: (1) Humana (1983; 1986); (2) Gastil (1987); (3) Bollen (1980); (4) Vanhanen (1984).

TABLE 6.6 Structure of government; levels and number of units in the 1980s

Aus.	Land (9); Gemeinde (2 300)
Belg.	Communauté (3); Provins (9); Arrondissement (44); Commune (596)
Den.	Amt (14); Kommuni (275)
Fin.	Lääni (12); Kommuni (461)
Fr.	Région (22); Département (96); Arrondissement (325); Canton (3 714); Commune (33 394)
Gr.	Nomos (52); Demos (6 023)
FRG	Land (11); Gemeinde (8 502)
Ice.	Region (8); Sýslur (23); Hreppur (216)
Ire.	Province (4); County/Corporation (38); District/Board (75)
It.	Regione (20); Provincia (94); Commune (8 075)
Lux.	Commune (118)
Neth.	Provins (12); Gemeente (714)
Norw.	Fylke (19); Kommune (454)
Port.	Region/Distrito (20); Concelho (305); Freguesia (4 050)
Sp.	Communidad Autónoma (17); Provincia (50); Municipio (8 056)
Sw.	Län (24); Kommun (284)
Switz.	Kanton (26); Kommune (3 029)
Turk.	İl (67); İlçe; Bucak
UK	England and Wales: County (53); District (369); Parishes/Communities(10 000/1 000) Scotland: Regions (9); Island Areas (3); Districts (53)
Can.	Province (12); Municipality (3 217)
USA	State (50); County (2 992); Municipality (78 200)
Jap.	Todofuken (47); Municipality (900)
Austral.	State (6); Municipality (900)
NZ.	Region (22); County/Municipality (224)

Source: The Statesman's Year-Book, 1989–90 (1989).

TABLE 6.7. Elected governments in Western Europe, 1945–1989

	Number of governments	Average duration (months)	Parliamentary support (average)	Number of parties in government (average)	Average government duration (%)
Aus.	17	30.8	76.4	1.8	64.3
Belg.	29	18.0	61.7	3.0	37.1
Den.	26	20.2	41.1	1.8	42.2
Fin.	38	13.6	57.4	2.9	28.3
Fr.	44	11.9	59.0	4.3	19.8
FRG	15	32.2	58.9	2.2	66.8
Gr.	39	11.0	63.3	1.1	22.9
Ice.	20	26.6	55.4	2.2	55.3
Ire.	18	31.8	50.9	1.6	65.3
It.	43	12.3	52.1	2.7	20.0
Lux.	12	45.1	69.3	2.0	73.6
Neth.	18	30.3	62.0	3.3	61.4
Norw.	22	25.2	47.6	1.8	52.5
Port.	14	11.5	60.0	1.6	23.8
Sp.	6	29.2	50.5	1.0	61.0
Sw.	22	24.7	47.2	1.4	60.9
Switz.	12	43.6	80.4	3.8	90.9
Turk.	27	20.5	64.1	1.7	41.3
UK	16	32.8	54.3	1.0	54.7
Can.	18	30.0	55.0	1.0	50.1
USA	14	40.1	49.0	1.0	83.5
Jap.	32	17.1	52.0	1.0	38.2
Austral.	24	22.4	59.1	1.7	62.1
NZ	19	30.4	57.2	1.0	83.7

Notes: A government is identified by means of its prime minister or president (USA) and the participating parties. If there is a change in either of these there is a new government. A change of government always follows an election, i.e. a government cannot stay longer than an election period. Parliamentary support refers to the support (in percentage of total seats) each government has had in the lower chambers of its parliament; number of parties in government refers to the number of parties participating in government; average government duration refers to the months a government holds office as a percentage of the maximum period of government. Greece did not hold democratic elections between 1967 and 1974. Portugal did not hold democratic elections in the post-war period until 1975. Spain did not hold democratic elections in the post-war period until 1977.

Sources: Paloheimo (1984); The *Europa Year Book* (various years); *Keesing's Contemporary Archives* (various years).

TABLE 6.8. Elected governments in Western Europe formed in the 1940s

	Number of governments	Average duration (months)	Parliamentary support (average)	Number of parties in government (average)	Average government duration
Aus.	3	29.3	95.0	2.3	61.0
Belg.	5	10.2	57.0	2.2	21.2
Den.	2	29.0	35.8	1.0	60.5
Fin.	3	19.7	64.8	3.3	41.0
Fr.	10	5.1	64.0	4.5	8.6
FRG	1	49.0	51.7	3.0	100.0
Gr.	6	6.7	74.6	2.0	14.0
Ice.	3	15.0	64.1	2.3	31.3
Ire.	2	42.5	50.4	3.0	88.5
It.	4	10.5	69.4	3.8	12.5
Lux.	1	52.0	65.4	2.0	87.0
Neth.	2	35.0	68.5	3.0	72.0
Norw.	2	36.0	53.7	1.0	75.0
Port.	—	—	—	—	—
Sp.	—	—	—	—	—
Sw.	3	24.7	49.6	1.0	51.3
Switz.	1	48.0	85.0	4.0	100.0
Turk.	3	19.0	84.9	1.0	39.7
UK	1	55.0	61.4	1.0	92.0
Can.	3	32.0	58.6	1.0	53.7
USA	2	44.5	53.5	1.0	92.5
Jap.	5	15.4	42.9	1.2	34.8
Austral.	3	23.3	61.7	1.3	63.7
NZ	3	31.0	54.6	1.0	83.3

Notes: A government is identified by means of its prime minister and the participating parties. If there is a change in either of these there is a new government. A change of government always follows an election, i.e. a government cannot stay longer than an election period. Parliamentary support refers to the support (in percentage of total seats) each government had in the lower chambers of its parliament; number of parties in government refers to the number of parties participating in government; average government duration refers to the months a government holds office as a percentage of an estimated maximum period of government. See notes to Table 6.7

Source: See Table 6.7.

TABLE 6.9. Elected governments in Western Europe formed in the 1950s

	Number of governments	Average duration (months)	Parliamentary support (average)	Number of parties in government (average)	Average government duration
Aus.	3	32.0	92.9	2.0	66.7
Belg.	6	21.5	52.2	1.3	44.2
Den.	6	18.8	43.0	1.7	39.0
Fin.	13	10.5	52.4	2.5	21.5
Fr.	17	8.6	55.0	5.5	14.3
FRG	2	49.0	63.1	3.0	100.0
Gr.	12	10.7	52.9	1.1	22.1
Ice.	6	27.3	47.6	1.8	56.8
Ire.	4	31.0	50.9	1.5	64.5
It.	10	12.0	48.9	1.8	20.0
Lux.	5	28.6	74.6	2.0	46.2
Neth.	4	32.5	70.0	3.8	66.0
Norw.	4	30.0	52.8	1.0	62.5
Port.	—	—	—	—	—
Sp.	—	—	—	—	—
Sw.	5	21.6	53.8	1.6	45.0
Switz.	4	36.0	74.5	3.5	75.0
Turk.	3	40.0	82.2	1.0	70.3
UK	5	30.4	53.9	1.0	50.6
Can.	3	31.3	61.8	1.0	52.3
USA	2	48.0	44.7	1.0	100.0
Jap.	8	11.5	47.3	1.0	26.3
Austral.	4	31.8	58.6	2.0	88.3
NZ	3	36.7	56.7	1.0	100.0

Notes: A government is identified by means of its prime minister and the participating parties. If there is a change in either of these there is a new government. A change of government always follows an election, i.e. a government cannot stay longer than an election period. Parliamentary support refers to the support (in percentage of total seats) each government had in the lower chambers of its parliament; number of parties in government refers to the number of parties participating in government; average government duration refers to the months a government holds office as a percentage of an estimated maximum period of government. See notes to Table 6.7.

Source: See Table 6.7.

TABLE 6.10. Elected governments in Western Europe formed in the 1960s

	Number of governments	Average duration (months)	Parliamentary support (average)	Number of parties in government (average)	Average government duration
Aus.	4	27.0	84.8	1.8	56.3
Belg.	4	32.0	67.7	2.0	60.3
Den.	6	22.5	48.5	2.0	47.0
Fin.	6	17.7	58.4	3.0	37.0
Fr.	5	24.2	66.9	2.6	40.4
FRG	5	26.6	64.9	2.0	55.4
Gr.	12	5.7	61.0	0.6	11.7
Ice.	2	40.0	53.3	2.0	83.5
Ire.	4	34.0	50.2	1.0	71.3
It.	8	15.0	48.4	1.9	25.1
Lux.	2	60.0	67.0	2.0	96.0
Neth.	4	24.0	58.0	3.3	48.5
Norw.	5	22.6	50.4	2.8	47.0
Port.	—	—	—	—	—
Sp.	—	—	—	—	—
Sw.	4	30.0	51.2	1.0	75.0
Switz.	2	48.0	85.0	4.0	100.0
Turk.	7	18.7	58.0	1.8	39.1
UK	3	26.7	55.3	1.0	44.3
Can.	5	25.2	50.0	1.0	42.0
USA	4	36.0	56.2	1.0	75.0
Jap.	6	24.0	60.4	1.0	50.8
Austral.	7	17.6	60.0	2.0	48.7
NZ	4	33.8	55.6	1.0	93.8

Notes: A government is identified by means of its prime minister and the participating parties. If there is a change in either of these there is a new government. A change of government always follows an election, i.e. a government cannot stay longer than an election period. Parliamentary support refers to the support (in percentage of total seats) each government had in the lower chambers of its parliament; number of parties in government refers to the number of parties participating in government; average government duration refers to the months a government holds office as a percentage of an estimated maximum period of government. See notes to Table 6.7.

Source: See Table 6.7.

TABLE 6.11. Elected governments in Western Europe formed in the 1970s

	Number of governments	Average duration (months)	Parliamentary support (average)	Number of parties in government (average)	Average government duration
Aus.	4	39.3	50.7	1.0	82.0
Belg.	7	13.6	66.6	4.3	28.1
Den.	7	17.4	36.1	1.1	36.3
Fin.	12	11.7	58.4	2.8	24.3
Fr.	5	21.2	59.3	3.4	35.4
FRG	3	31.7	53.5	2.0	66.7
Gr.	4	17.5	73.8	1.0	36.3
Ice.	5	23.0	52.0	2.2	47.8
Ire.	3	33.0	54.8	1.3	66.0
It.	11	11.2	49.6	2.6	18.5
Lux.	2	60.5	60.4	2.0	100.0
Neth.	4	30.5	55.4	4.3	59.8
Norw.	5	23.8	41.9	1.4	49.6
Port.	8	7.3	35.1	1.3	15.1
Sp.	2	21.5	47.6	1.0	45.0
Sw.	5	25.4	40.8	1.8	70.6
Switz.	3	48.0	83.3	4.0	100.0
Turk.	10	11.4	57.1	2.3	23.8
UK	5	31.4	50.7	1.0	52.4
Can.	3	30.0	47.6	1.0	50.0
USA	3	32.0	47.6	1.0	66.7
Jap.	6	18.0	52.6	1.0	41.0
Austral.	6	19.3	57.9	1.7	53.7
NZ	5	23.6	59.7	1.0	65.6

Notes: A government is identified by means of its prime minister and the participating parties. If there is a change in either of these there is a new government. A change of government always follows an election, i.e. a government cannot stay longer than an election period. Parliamentary support refers to the support (in percentage of total seats) each government had in the lower chambers of its parliament; number of parties in government refers to the number of parties participating in government; average government duration refers to the months a government holds office as a percentage of the maximum period of government.

Source: See Table 6.7.

TABLE 6.12. Elected governments in Western Europe formed in the 1980s

	Number of governments	Average duration (months)	Parliamentary support (average)	Number of parties in government (average)	Average government duration
Aus.	3	22.0	64.3	2.0	46.0
Belg.	7	16.8	64.6	4.3	35.0
Den.	5	19.5	38.8	3.2	40.8
Fin.	4	21.0	60.6	3.8	43.7
Fr.	7	14.2	55.7	3.0	23.7
FRG	4	25.3	55.2	2.0	52.7
Gr.	5	28.3	65.1	1.6	58.8
Ice.	4	33.7	65.8	2.8	70.0
Ire.	5	24.0	49.5	1.6	48.0
It.	10	12.3	54.3	4.0	20.4
Lux.	2	60.0	69.3	2.0	100.0
Neth.	4	32.3	61.2	2.3	67.3
Norw.	6	21.0	44.6	2.0	43.8
Port.	6	18.2	53.2	1.9	37.6
Sp.	4	34.3	52.0	1.0	71.7
Sw.	5	22.3	42.5	1.2	61.8
Switz.	2	48.0	81.3	4.0	100.0
Turk.	4	37.0	60.9	1.0	77.0
UK	2	48.0	59.5	1.0	80.0
Can.	4	34.7	59.2	1.0	57.7
USA	3	48.0	40.6	1.0	100.0
Jap.	7	18.2	56.4	1.0	41.3
Austral.	4	26.7	58.2	1.3	74.0
NZ	4	30.7	57.8	1.0	84.3

Notes: A government is identified by means of its prime minister and the participating parties. If there is a change in either of these there is a new government. A change of government always follows an election, i.e. a government cannot stay longer than an election period. Parliamentary support refers to the support (in percentage of total seats) each government had in the lower chambers of its parliament; number of parties in government refers to the number of parties participating in government; average government duration refers to the months a government holds office as a percentage of an estimated maximum period of government.

Source: See Table 6.7.

Section 7. Political Parties and Elections

TABLES

7.1. National electoral participation: valid votes as a percentage of the electorate
7.2. National electoral participation: total votes as a percentage of the electorate
7.3. Classification of political parties, by country
7.4. Electoral strength of communist parties in national elections
7.5a. Electoral strength of religious parties in national elections (CDU/CSU, ÖVP, DC included)
7.5b. Electoral strength of religious parties in national elections (CDU/CSU, ÖVP, DC excluded)
7.6. Electoral strength of socialist parties in national elections
7.7. Electoral strength of ethnic parties in national elections
7.8. Electoral strength of agrarian parties in national elections
7.9. Electoral strength of left/socialist parties in national elections
7.10. Electoral strength of liberal parties in national elections
7.11a. Electoral strength of conservative parties in national elections (CDU/CSU, ÖVP, DC excluded)
7.11b. Electoral strength of conservative parties in national elections (CDU/CSU, ÖVP, DC included)
7.12. Electoral strength of protest parties in national elections
7.13. Electoral strength of ultra-right parties in national elections
7.14. Electoral strength of environmentalist/green parties in national elections
7.15. Electoral strength of other parties in national elections

TABLE 7.1. National electoral participation: valid votes as a percentage of the electorate

	1945–9	1950–4	1955–9	1960–4	1965–9	1970–4	1975–9	1980–4	1985–9
Aus.	94.4	94.2	93.6	92.7	92.7	91.2	91.6	91.3	88.8
Belg.	88.1	87.9	89.0	87.2	84.5	83.7	87.7	86.1	86.9
Den.	85.8	80.8	83.3	85.4	88.7	87.4	86.7	85.3	85.6
Fin.	76.1	76.8	74.6	84.8	84.6	81.5	74.3	75.4	71.7
Fr.	78.4	78.0	77.2	66.6	78.9	79.4	81.7	69.9	69.6
FRG	76.0	83.0	84.5	84.3	85.0	90.4	89.9	88.1	83.5
Gr.	—	75.9	74.9	81.7	—	78.6	80.1	77.5	82.9
Ice.	86.9	88.4	89.5	89.5	89.7	89.7	87.9	86.4	88.5
Ire.	73.5	75.2	70.6	69.9	75.3	75.7	75.7	73.7	70.2
It.	86.2	89.5	91.1	89.9	89.4	90.1	88.8	84.0	84.6
Lux.	86.7	92.6	87.9	85.1	83.2	85.2	82.7	83.4	87.3
Neth.	90.5	92.1	93.4	92.7	92.3	80.7	87.5	83.6	82.7
Norw.	78.6	78.9	77.9	78.6	84.4	80.1	82.8	81.9	83.6
Port.	—	—	—	—	—	—	84.0	79.2	70.7
Sp.	—	—	—	—	—	—	71.3	78.9	69.1
Sw.	82.4	78.7	78.1	84.5	88.7	89.4	90.8	90.6	87.0
Switz.	70.3	68.4	67.3	63.3	62.6	56.6	49.6	48.2	46.1
Turk.	—	—	—	—	—	—	—	—	88.9
UK	72.6	82.8	77.8	77.1	75.8	74.5	76.0	72.7	75.3
Can.	74.8	67.1	77.0	79.5	75.0	71.9	75.1	72.0	74.9
USA	53.3	63.7	61.6	64.4	62.3	57.1	55.8	55.1	52.8
Jap.	70.1	74.6	75.8	71.6	70.4	71.1	70.0	70.2	69.9
Austral.	92.9	90.7	88.9	93.4	92.4	93.5	93.1	90.9	89.2
NZ	95.1	89.7	93.0	89.9	87.2	88.5	82.6	90.0	86.6

Sources: Mackie and Rose (1982); *Keesing's Record of World Events* (various years); *European Journal of Political Research* (various years).

TABLE 7.2. National electoral participation: total votes as a percentage of the electorate

	1945–9	1950–4	1955–9	1960–4	1965–9	1970–4	1975–9	1980–4	1985–9
Aus.	95.6	95.8	95.1	93.8	93.8	92.1	92.6	92.6	90.4
Belg.	92.4	92.9	93.6	92.3	90.8	90.5	95.0	94.6	93.5
Den.	86.1	81.1	83.7	85.6	89.0	88.0	87.2	85.8	86.3
Fin.	76.4	77.3	75.0	85.1	84.9	81.8	74.6	75.7	72.1
Fr.	79.9	80.2	79.6	68.8	80.5	81.2	83.3	70.9	71.9
FRG	78.5	85.8	87.8	87.7	86.8	91.1	90.7	88.9	84.3
Gr.	—	76.3	75.2	82.2	—	79.5	81.1	78.6	84.2
Ice.	88.2	89.9	91.0	91.1	91.4	90.9	89.8	88.9	89.5
Ire.	74.2	75.9	71.3	70.6	76.0	76.6	76.3	74.4	70.9
It.	90.7	93.9	93.7	92.9	92.8	93.1	91.9	89.0	90.5
Lux.	91.3	92.6	92.3	90.6	88.6	90.1	88.9	88.8	92.1
Neth.	93.4	95.0	95.5	95.1	94.9	81.3	88.0	84.1	82.9
Norw.	79.2	79.3	78.3	79.1	84.6	80.2	82.9	82.0	83.7
Port.	—	—	—	—	—	—	88.3	81.2	72.4
Sp.	—	—	—	—	—	—	72.5	79.5	70.4
Sw.	82.7	79.1	78.5	84.9	89.3	89.6	91.3	91.4	88.0
Switz.	71.7	69.8	68.6	64.5	63.8	56.8	50.4	48.9	46.8
Turk.	—	84.8	89.8	81.0	67.8	66.8	72.5	92.3	91.3
UK	—	—	—	77.2	76.0	74.7	76.3	72.8	75.4
Can.	75.6	67.9	77.8	80.2	75.8	74.1	75.8	72.5	75.5
USA	—	—	—	—	—	—	—	—	—
Jap.	71.4	75.3	76.5	72.3	71.3	71.8	70.7	71.4	71.4
Austral.	95.0	92.2	91.5	95.5	95.1	95.4	95.2	94.4	93.8
NZ	95.8	90.2	93.4	90.4	87.8	89.1	83.1	91.2	89.0

Source: See Table 7.1.

TABLE 7.3 Classification of political parties, by country

Party type	Party
Austria	
Communist	Kommunistische Partei Österreichs
Environmental (Green)	Vereinte Grüne Österreichs
Environmental (Green)	Alternative Liste Österreichs
Liberal	Freiheitliche Partei Österreichs
Protest	Demokratische Fortschrittliche Partei
Religious	Österreichische Volkspartei
Socialist	Sozialistische Partei Österreichs
Belgium	
Communist	Kommunistische Partij van België/Parti Communiste de Belgique
Environmental (Green)	AGALEV
Environmental (Green)	Écologistes
Ethnic	Christelijk Vlaamse Volksunie
Ethnic	Partei der Deutschsprächigen Belgier
Ethnic	Front Démocratique des Bruxellois Francophones
Ethnic	Front Démocratique Wallon
Ethnic	Front Wallon
Ethnic	Parti Wallon des Travailleurs
Ethnic	Rassemblement Wallon
Ethnic	Vlaams Blok
Left-Socialist	Parti du Travail de Belgique
Left-Socialist	Parti Ouvrier Socialiste
Liberal	Partij voor Vrijheid en Vooruitgang/Parti de la Liberté et du Progrès
Liberal	Partij voor Vrijheid en Vooruitgang
Liberal	Parti Réformateur Libéral
Liberal	Parti Libéral
Protest	Union Démocratique pour le Respect du Travail/Respect voor Arbeid en Demokratie
Religious	Parti Social Chrétien/Christelijke Volkspartij
Religious	Dissident Catholic Lists
Religious	Union Démocratique Belge
Religious	Christelijke Volkspartij
Religious	Parti Social Chrétien
Socialist	Belgische Socialistische Partij/Parti Socialiste Belge
Socialist	Belgische Socialistische Partij
Socialist	Parti Socialiste Belge

Comparative Tables

Denmark

Agrarian	Venstre
Communist	Danmarks Kommunistiske Parti
Conservative	Konservative Folkeparti
Conservative	Uafhængige
Environmental (Green)	Grøne
Ethnic	Slesvigsk Parti/Schleswigsche Partei
Left-Socialist	Socialistisk Folkeparti
Left-Socialist	Venstresocialisterne
Left-Socialist	Fælles Kurs
Liberal	Radikale Venstre
Liberal	Retsforbundet
Liberal	Dansk Samling
Liberal	Liberalt Centrum
Liberal	Centrum-Demokraterne
Protest	Fremskridtspartiet
Religious	Kristeligt Folkeparti
Socialist	Socialdemokratiet

Finland

Agrarian	Keskustapuolue
Agrarian	Suomen Pienviljelijäin Puolue
Communist	Suomen Kansan Demokraattinen Liitto
Communist	Demokraattinen Vaihtoehto
Conservative	Kansallinen Kokomos
Conservative	Suomen Perustuslaillinen Kansanpuolue
Environmental (Green)	Green
Ethnic	Svenska Folkpartiet
Left-Socialist	Työväen ja Pienviljelijäin Sosialdemokraattinen Liitto
Liberal	Liberaalinen Kansanpuolue
Liberal	Vapaamielisten Liitto
Protest	Suomen Maaseudun Puolue
Protest	Suomen Kansan Yhtenäisyyden Puolue
Protest	Suomen Eläkeläisten Puolue
Religious	Suomen Kristillinen Liitto
Socialist	Suomen Sosialdemokraattinen Puolue

France

Communist	Parti Communiste Français
Conservative	Centre National des Indépendants
Conservative	Gaullistes
Conservative	Parti Républicain
Conservative	Centre Démocratie et Progrès
Environmental (Green)	Écologistes
Left-Socialist	Union des Forces Démocratiques
Left-Socialist	Parti Socialiste Unifié

Liberal	Parti Républicain Radical et Radical Socialiste
Protest	Union pour la Défense des Commerçants et Artisans
Religious	Mouvement Républicain Populaire
Religious	Centre du Progrès et de la Démocratie Moderne
Religious	Mouvement Réformateur
Religious	Centre des Démocrates Sociaux
Socialist	Parti Socialiste
Ultra-Right	Front National

Federal Republic of Germany

Communist	Kommunistische Partei Deutschlands
Communist	Deutsche Friedensunion
Communist	Aktion Demokratischer Fortschritt
Conservative	Deutsche Partei
Conservative	Gesamtdeutsche Partei
Environmental (Green)	Die Grünen
Ethnic	Bayernpartei
Ethnic	Südschleswiger Wählerverband
Ethnic	Gesamtdeutscher Block/Bund der Heimatvertriebenen und Entrechteten
Ethnic	Föderalistische Union
Liberal	Freie Demokratische Partei
Protest	Wirtschaftliche Aufbauvereinigung
Religious	Zentrumspartei
Religious	Christlich Demokratische Union/Christlich Soziale Union
Socialist	Sozialdemokratische Partei Deutschlands
Socialist	Gesamtdeutsche Volkspartei
Ultra-Right	Deutsche Reichspartei
Ultra-Right	Nationaldemokratische Partei Deutschlands

Greece

Communist	Communist Party of Greece (KKE)
Communist	United Democratic Left (EDA)
Communist	Christian Democracy (CD)
Communist	Communist Party of the Interior (KKEes)
Conservative	People's Party (LK)
Conservative	National Unity Party (EEK)
Conservative	Reformist Party (MK)
Conservative	Nationalist Party (KE)
Conservative	New Party (NK)
Conservative	Greek Rally (ES)
Conservative	National Radical Union (ERE)
Conservative	Popular Social Party (LKK)
Conservative	National Democratic Union (EDE)
Conservative	New Democracy (ND)

Conservative	New Liberal Party (KN)
Conservative	Party of the Progressives (KP)
Conservative	Democratic Renewal (DIANA)
Environmental (Green)	Ecologists
Ethnic	Independent Muslims
Liberal	Liberal Party (KF)
Liberal	Republican Socialist Party (DSK)
Liberal	National Party of Greece (EKE)
Liberal	National Progressive Centre Union (EPEK)
Liberal	Farmers' and Workers' Rally (SAE)
Liberal	Democratic Party of the Working People (DKEL)
Liberal	Progressive Party (PK)
Liberal	Union of the Centre (EDHIK)
Liberal	Liberal Party (KF)
Liberal	Party for Democratic Socialism (KODISO)
Socialist	Pan-Hellenic Socialist Movement (PASOK)
Ultra-Right	National Front (EM)

Iceland

Agrarian	Progressive Party
Agrarian	National Party
Agrarian	Association for Equality and Justice
Communist	Communist Party
Conservative	Independence Party (II)
Conservative	Republic Party
Conservative	Independent Democratic Party
Conservative	Independent Party
Environmental (Green)	Women's List
Environmental (Green)	Humanist Party
Left-Socialist	National Preservation Party
Liberal	Union of Liberals and Leftists
Protest	Citizens' Party
Socialist	Social Democrats
Socialist	Social Democratic Federation
Socialist	Social Democratic Alliance

Ireland

Agrarian	Clann na Poblachta
Environmental (Green)	Green Alliance
Ethnic	Sinn Féin (II)
Ethnic	National Progressive Democrats
Ethnic	National H-Block Committee
Left-Socialist	The Workers' Party
Left-Socialist	Democratic Socialist Party
Liberal	Fianna Fáil
Liberal	Clann na Talmhan
Liberal	Progressive Democrats

Religious	Fine Gael
Socialist	Irish Labour Party
Socialist	National Labour

Italy

Communist	Partito Comunista Italiano
Conservative	Partito Monarchico Popolare
Environmental (Green)	Partito Radicale
Environmental (Green)	Lista Verde
Ethnic	Partito Sardo d'Azione
Ethnic	Movimento per l'Independenciza della Sicilia
Ethnic	Südtiroler Volkspartei
Ethnic	Union Valdotaine
Ethnic	Lista per Trieste
Ethnic	Liga Veneta
Left-Socialist	Partito Socialista Italiano di Unità Proletaria
Left-Socialist	Manifesto/Partito di Unità Proletaria per il Comunismo
Liberal	Partito Repubblicano Italiano
Liberal	Partito d'Azione
Liberal	Partito Liberale Italiano
Protest	Fronte dell'Uomo Qualunque
Protest	Communità
Protest	Partito Nazionale dei Pensionati
Religious	Democrazia Cristiana
Socialist	Partito Socialista Italiano
Socialist	Partito Socialista Democratico Italiano
Socialist	Partito Socialista Unificato
Ultra-Right	Movimento Sociale Italiano-Destra Nazionale

Luxembourg

Communist	Parti Communiste Luxembourgeois
Environmental (Green)	Écologistes
Ethnic	Indépendants de l'Est
Liberal	Parti Démocratique
Protest	Parti des Classes Moyennes
Protest	Mouvement Indépendent Populaire
Protest	Enrôlés de Force
Protest	Aktiounskomitee
Religious	Parti Social-Chrétien
Socialist	Parti Ouvrier Socialiste Luxembourgeois
Socialist	Parti Social-Démocrate
Socialist	Socialistes Indépendants

The Netherlands

Communist	Communistische Partij Nederland
Environmental (Green)	Federatieve Groenen
Ethnic	Centrumpartij

Left-Socialist	Pacifistisch-Socialistische Partij
Left-Socialist	Socialistische Partij
Liberal	Volkspartij voor Vrijheid en Democratie
Liberal	Democraten '66
Protest	Boerenpartij
Protest	Middenstands Partij
Religious	Anti-Revolutionaire Partij
Religious	Katholieke Volkspartij
Religious	Christelijk-Historische Unie
Religious	Staatkundig Gereformeerde Partij
Religious	Katholieke Nationale Partij
Religious	Gereformeerd Politiek Verbond
Religious	Politieke Partij Radicalen
Religious	Rooms-Katholieke Partij Nederland
Religious	Christen Democratisch Appel
Religious	Reformatorische Politieke Federatie
Religious	Evangelische Volkspartij
Socialist	Partij van der Arbeid
Socialist	Democratische Socialisten '70

Norway

Agrarian	Senterpartiet
Communist	Norges Kommunistiske Parti
Conservative	Høyre
Conservative	Samfunnspartiet
Environmental (Green)	Green Party
Left-Socialist	Sosialistisk Venstreparti
Liberal	Venstre
Liberal	Liberale Folkpartiet
Protest	Fremskrittspartiet
Religious	Kristeligt Folkeparti
Socialist	Det Norske Arbeiderparti

Portugal

Communist	Partido Comunista Português
Communist	Movimento Democrático Português
Conservative	Partido Popular Monárquico
Environmental (Green)	Partido 'Os Verdes'
Left-Socialist	Movimento de Esquerda Socialista
Left-Socialist	União Democrático Popular
Left-Socialist	Frente Socialista Popular
Left-Socialist	Partido Socialista Revolucionário
Left-Socialist	Partido Operário de Unidade Socialista
Liberal	Partido Social Democrata
Liberal	Partido Renovador Democrático
Religious	Partido do Centro Democrático Social
Religious	Partido da Democracia Cristã

Socialist	Partido Socialista Português
Socialist	Uniao de Esquerda Democrática Socialista

Spain

Communist	Partido Comunista de España
Conservative	Unión del Centro Democrático
Conservative	Alianza Popular
Conservative	Centro Democrático y Social
Environmental (Green)	Los Verdes
Ethnic	Partido Nacionalista Vasco
Ethnic	Partido Carlista
Ethnic	Partido Socialista de Andalucía
Ethnic	Unió Democràtica de Catalunya
Ethnic	Convergència Democràtica de Catalunya
Ethnic	Euzkadiko Ezkerra
Ethnic	Bloque Nacional Popular Gallego
Ethnic	Partido Socialista Gallego
Ethnic	Herri Batasuna
Ethnic	Partido Aragonés Regionalista
Ethnic	Coalición Unión del Pueblo Canario
Ethnic	Unión Valenciana
Ethnic	Esquerra Republicana de Catalunya
Left-Socialist	Partido del Trabajo de España
Religious	Equipo de la Democracia Cristiana
Socialist	Partido Socialista Obrero Español
Socialist	Partido Socialista Popular
Ultra-Right	Alianza Nacional del 18 de Julio
Ultra-Right	Fuerza Nueva
Ultra-Right	Falange Española de la JONS

Sweden

Agrarian	Centerpartiet
Communist	Vänsterpartiet
Conservative	Moderata Samlingspartiet
Environmental (Green)	Miljöpartiet
Liberal	Folkpartiet
Religious	Kristdemokratiska Samhällspartiet
Socialist	Socialdemokratiska Arbetarpartiet

Switzerland

Agrarian	Schweizerische Volkspartei
Communist	Partei der Arbeit der Schweiz
Environmental (Green)	Grüne
Ethnic	Entente jurassienne
Left-Socialist	Partito Socialista Autonomo
Left-Socialist	Progressive Organisationen der Schweiz
Liberal	Demokraten
Liberal	Liberale-konservative Partei

Liberal	Freisinnige-demokratische Partei
Liberal	Freiwirtschaftler
Liberal	Landesring der Unabhängigen
Protest	Nationale Aktion gegen die Überfremdung von Volk und Heimat
Protest	Schweizerische Republikanische Bewegung
Protest	Schweizer Auto-Partei
Religious	Christlich Demokratische Volkspartei
Religious	Evangelische Volkspartei
Socialist	Sozialdemokratische Partei der Schweiz

Turkey

Conservative	Justice Party
Conservative	Republican Reliance Party
Conservative	Motherland Party (ANAP)
Conservative	True Path Party (TPP)
Left-Socialist	Turkish Labour Party
Liberal	Turkish Unity Party
Liberal	Reformist Democracy Party
Religious	National Salvation Party
Religious	Prosperity Party (Welfare Party)
Religious	Nationalist Labour Party (NLP)
Socialist	Republican People's Party
Socialist	Social Democratic Populist Party (SDPP)
Socialist	Democratic Left Party (DLP)
Ultra-Right	National Action Party
Ultra-Right	Democratic Party
Ultra-Right	New Turkey Party
Ultra-Right	Nation Party
Ultra-Right	Nationalist Democrat Party

United Kingdom

Communist	Communist Party
Conservative	Conservative Party
Conservative	National Liberal Party
Environmental (Green)	Ecology Party
Ethnic	United Ireland
Ethnic	Scottish National Party
Ethnic	Plaid Cymru
Ethnic	Social Democratic and Labour Party
Ethnic	Ulster Unionists and Loyalists
Liberal	Liberal Party
Socialist	Independent Labour Party
Socialist	Labour Party
Socialist	Social Democratic Party
Ultra-Right	National Front

Canada
Communist	Communist Party
Conservative	Progressive Conservative Party
Environmental (Green)	Green Party
Ethnic	Parti Nationaliste du Québec
Ethnic	Confederation of Regions—Western Party
Liberal	Liberal Party
Liberal	Social Credit
Liberal	Ralliement des Créditistes du Québec
Liberal	Libertarian Party
Protest	Rhinoceros Party
Socialist	New Democratic Party
Socialist	Bloc Populaire Canadien

USA
Communist	Communist Party
Conservative	Republicans
Conservative	Independent
Liberal	Democrats
Liberal	Prohibition
Liberal	States Rights
Liberal	Libertarian Party
Protest	American Independent Party
Socialist	Socialist Labor Party
Socialist	Socialist Party
Socialist	Progressive

Japan
Communist	Communist Party
Conservative	Democratic Liberal Party
Conservative	Hatoyama Liberal Party
Conservative	Yoshida Liberal Party
Conservative	Democratic Party
Conservative	Liberal Democratic Party
Liberal	Progressive Party
Liberal	Co-operative Party
Liberal	Social Reform Party
Liberal	New Liberal Club
Liberal	Social Democratic Federation
Religious	Clean Government Party (Komeito)
Socialist	Socialist Party
Socialist	Labour-Farmer Party
Socialist	Left-Wing Socialist Party
Socialist	Right-Wing Socialist Party
Socialist	Democratic Socialist Party

Australia
Agrarian	Country Party

Communist	Communist Party
Conservative	Australian Liberal Party
Conservative	Country-Liberal Party
Liberal	Services Party of Australia
Liberal	Australia Party
Liberal	Liberal Movement
Liberal	National Alliance
Liberal	Australian Democrats
Socialist	Australian Labor Party
Socialist	Lang Labor Party
Socialist	Democratic Labor Party
Socialist	Queensland Labor Party

New Zealand

Communist	Communist Party
Conservative	National Party
Environmental (Green)	Values Party
Ethnic	Manu Motuhake
Liberal	Social Credit Political League = Democrats
Religious	New Zealand Party
Socialist	Labour Party

TABLE 7.4. Electoral strength of communist parties in national elections

	1945–9	1950–4	1955–9	1960–4	1965–9	1970–4	1975–9	1980–4	1985–9
Aus.	5.3	5.3	3.9	3.0	0.4	1.2	1.1	0.7	0.7
Belg.	10.1	4.2	1.9	3.1	4.0	3.2	2.7	2.3	1.0
Den.	9.7	4.6	3.1	1.2	0.9	2.5	3.3	0.9	0.9
Fin.	21.8	21.6	23.2	22.0	21.2	16.8	18.4	13.8	13.9
Fr.	27.0	26.7	22.6	21.8	21.3	21.4	20.6	16.1	10.6
FRG	5.7	2.2	—	1.9	1.0	0.3	0.3	0.2	0.0
Gr.	—	10.0	17.3	13.6	—	9.5	12.1	12.3	12.1
Ice.	19.5	16.0	16.8	16.0	13.9	17.7	21.3	17.3	13.3
Ire.	—	—	—	—	—	—	—	—	—
It.	20.6	22.6	22.7	25.3	26.9	27.2	32.4	29.9	26.6
Lux.	11.8	8.9	9.1	12.5	15.5	10.5	5.8	5.0	5.1
Neth.	9.2	6.2	3.6	2.8	3.6	4.2	1.7	2.0	0.7
Norw.	8.9	5.1	3.4	2.9	1.2	—	0.4	0.3	0.2
Port.	—	—	—	—	—	—	17.6	18.0	14.3
Sp.	—	—	—	—	—	—	10.1	4.1	7.4
Sw.	6.3	4.3	4.2	4.9	3.0	5.1	5.2	5.6	5.6
Switz.	5.1	2.7	2.7	2.2	2.9	2.6	2.3	0.9	0.8
Turk.	—	—	—	—	—	—	—	—	—
UK	0.4	0.2	0.1	0.2	0.2	0.1	0.1	0.0	0.0
Can.	1.4	1.1	0.1	0.1	0.1	0.1	0.1	0.1	—
USA	0.0	0.0	0.0	0.0	0.0	0.0	0.1	0.0	—
Jap.	5.8	2.2	2.3	3.5	5.8	10.5	10.4	9.6	8.8
Austral.	1.2	1.1	0.9	0.6	0.3	0.1	0.2	0.0	0.0
NZ	0.2	0.1	0.1	0.3	0.1	0.0	0.0	0.0	0.0

Source: See Table 7.1.

TABLE 7.5a. Electoral strength of religious parties in national elections (CDU/CSU, ÖVP, DC included)

	1945–9	1950–4	1955–9	1960–4	1965–9	1970–4	1975–9	1980–4	1985–9
Aus.	46.9	41.3	45.1	45.4	48.3	43.9	42.4	43.0	41.3
Belg.	44.2	44.9	46.5	44.4	33.3	31.3	36.1	26.4	28.4
Den.	—	—	—	—	—	3.0	3.8	2.5	2.2
Fin.	—	—	0.2	0.8	0.4	1.8	4.1	3.0	2.6
Fr.	26.4	12.5	11.2	8.9	11.5	16.2	5.3	5.2	—
FRG	34.1	46.0	50.2	45.3	46.9	44.9	48.6	46.7	44.3
Gr.	—	—	—	—	—	—	—	—	—
Ice.	—	—	—	—	—	—	—	—	—
Ire.	19.8	28.9	26.6	32.0	34.1	35.1	30.5	37.7	28.2
It.	41.9	40.1	42.4	38.2	39.0	38.7	38.5	32.9	34.3
Lux.	39.2	42.4	36.9	33.3	35.3	27.9	34.5	34.9	31.7
Neth.	55.4	54.7	52.5	52.2	47.4	41.9	37.8	36.7	40.5
Norw.	8.2	10.5	10.2	9.6	8.8	12.3	12.4	9.4	8.4
Port.	—	—	—	—	—	—	14.3	22.3	8.0
Sp.	—	—	—	—	—	—	0.7	—	—
Sw.	—	—	—	0.9	1.5	1.8	1.4	1.9	2.7
Switz.	22.1	23.5	24.5	25.0	23.7	22.8	23.4	22.5	20.0
Turk.	—	—	—	—	—	11.9	8.6	—	7.2
UK	—	—	—	—	—	—	—	—	—
Can.	—	—	—	—	—	—	—	—	—
USA	—	—	—	—	—	—	—	—	—
Jap.	—	—	—	—	8.2	8.5	10.4	9.6	9.4
Austral.	—	—	—	—	—	—	—	—	—
NZ	—	—	—	—	—	—	—	6.2	0.3

Source: See Table 7.1.

TABLE 7.5b. Electoral strength of religious parties in national elections (CDU/CSU, ÖVP, DC excluded)

	1945–9	1950–4	1955–9	1960–4	1965–9	1970–4	1975–9	1980–4	1985–9
Aus.	—	—	—	—	—	—	—	—	—
Belg.	44.2	44.9	46.5	44.4	33.3	31.3	36.1	26.4	28.4
Den.	—	—	—	—	—	3.0	3.8	2.5	2.2
Fin.	—	—	0.2	0.8	0.4	1.8	4.1	3.0	2.6
Fr.	26.4	12.5	11.2	8.9	11.5	16.2	5.3	5.2	—
FRG	—	—	—	—	—	—	—	—	—
Gr.	—	—	—	—	—	—	—	—	—
Ice.	—	—	—	—	—	—	—	—	—
Ire.	19.8	28.9	26.6	32.0	34.1	35.1	30.5	37.7	28.2
It.	—	—	—	—	—	—	—	—	—
Lux.	39.2	42.4	36.9	33.3	35.3	27.9	34.5	34.9	31.7
Neth.	55.4	54.7	52.5	52.2	47.4	41.9	37.8	36.7	40.5
Norw.	8.2	10.5	10.2	9.6	8.8	12.3	12.4	9.4	8.4
Port.	—	—	—	—	—	—	14.3	22.3	8.0
Sp.	—	—	—	—	—	—	0.7	—	—
Sw.	—	—	—	0.9	1.5	1.8	1.4	1.9	2.7
Switz.	22.1	23.5	24.5	25.0	23.7	22.8	23.4	22.5	20.0
Turk.	—	—	—	—	—	11.9	8.6	—	7.2
UK	—	—	—	—	—	—	—	—	—
Can.	—	—	—	—	—	—	—	—	—
USA	—	—	—	—	—	—	—	—	—
Jap.	—	—	—	—	8.2	8.5	10.4	9.6	9.4
Austral.	—	—	—	—	—	—	—	—	—
NZ	—	—	—	—	—	—	—	6.2	0.3

Source: See Table 7.1.

TABLE 7.6. Electoral strength of socialist parties in national elections

	1945–9	1950–4	1955–9	1960–4	1965–9	1970–4	1975–9	1980–4	1985–9
Aus.	41.7	42.1	43.9	44.0	42.6	49.2	50.7	47.7	43.1
Belg.	31.0	37.0	37.0	36.7	28.1	27.0	26.3	25.2	29.5
Den.	36.4	40.4	39.4	42.0	36.2	31.5	35.1	32.3	29.6
Fin.	25.7	26.4	23.2	19.5	27.2	24.6	24.4	26.7	24.1
Fr.	20.9	14.5	15.3	12.7	17.7	19.7	25.0	37.8	35.2
FRG	29.2	30.0	31.8	36.2	41.0	45.8	42.6	40.6	37.0
Gr.	—	—	—	—	—	13.6	25.3	48.1	41.9
Ice.	17.2	15.6	15.3	14.2	15.7	9.8	19.7	19.0	15.4
Ire.	11.3	11.8	9.1	12.0	16.2	13.7	11.6	9.5	7.9
It.	18.3	17.2	18.8	19.9	14.5	14.7	13.3	15.5	17.3
Lux.	32.4	35.1	34.9	37.7	32.3	38.3	32.5	36.1	27.2
Neth.	27.0	29.0	31.6	28.0	23.6	30.7	34.5	29.7	32.6
Norw.	43.4	46.7	48.3	46.8	44.8	35.3	42.3	37.2	37.6
Port.	—	—	—	—	—	—	35.4	32.9	22.1
Sp.	—	—	—	—	—	—	32.7	46.5	42.3
Sw.	46.1	46.0	45.4	47.6	50.1	44.5	43.0	45.6	44.0
Switz.	26.2	26.0	26.7	26.6	23.5	22.9	24.7	22.9	18.4
Turk.	—	37.7	40.9	36.7	28.1	33.2	41.4	30.5	33.2
UK	48.2	47.5	45.1	44.1	48.0	39.8	36.9	39.3	40.5
Can.	16.3	11.3	10.1	13.3	17.5	16.6	17.9	19.3	20.4
USA	2.8	0.2	0.1	0.1	0.1	0.1	—	—	—
Jap.	19.8	24.8	31.6	36.2	32.2	28.9	26.8	26.4	23.6
Austral.	49.0	45.3	51.0	54.8	50.2	52.8	42.6	47.7	45.8
NZ	49.3	45.0	48.3	43.6	42.8	48.4	40.0	41.0	48.0

Source: See Table 7.1.

TABLE 7.7. Electoral strength of ethnic parties in national elections

	1945–9	1950–4	1955–9	1960–4	1965–9	1970–4	1975–9	1980–4	1985–9
Aus.	—	—	—	—	—	—	—	—	—
Belg.	1.1	1.1	2.0	3.5	12.3	19.0	16.4	15.3	11.1
Den.	0.2	0.4	0.4	0.4	0.1	0.1	—	—	—
Fin.	8.1	7.3	6.7	6.4	6.0	5.6	4.8	4.9	5.6
Fr.	—	—	—	—	—	—	—	—	—
FRG	4.5	7.8	5.6	0.1	—	—	—	—	—
Gr.	—	—	—	—	—	—	—	—	—
Ice.	—	—	—	—	—	—	—	—	0.4
Ire.	—	—	5.3	2.1	—	—	—	—	1.5
It.	0.9	0.6	0.6	0.5	0.7	0.6	0.8	1.3	2.7
Lux.	0.6	—	—	—	—	—	—	—	—
Neth.	—	—	—	—	—	—	—	—	—
Norw.	—	—	—	—	—	—	—	—	—
Port.	—	—	—	—	—	—	—	—	—
Sp.	—	—	—	—	—	—	8.4	8.7	10.9
Sw.	—	—	—	—	—	—	—	—	—
Switz.	—	—	—	—	—	—	0.5	—	—
Turk.	—	—	—	—	—	4.3	—	—	—
UK	0.6	0.5	0.7	0.9	1.2	—	3.9	3.7	3.7
Can.	—	—	—	—	—	—	—	0.6	0.3
USA	—	—	—	—	—	—	—	—	—
Jap.	—	—	—	—	—	—	—	—	—
Austral.	—	—	—	—	—	—	—	—	—
NZ	—	—	—	—	—	—	—	0.2	0.5

Source: See Table 7.1.

TABLE 7.8. Electoral strength of agrarian parties in national elections

	1945–9	1950–4	1955–9	1960–4	1965–9	1970–4	1975–9	1980–4	1985–9
Aus.	—	—	—	—	—	—	—	—	—
Belg.	—	—	—	—	—	—	—	—	—
Den.	25.5	22.2	25.1	21.0	19.0	14.0	15.9	11.7	11.2
Fin.	23.6	23.9	23.1	23.0	21.2	16.8	17.5	17.6	17.6
Fr.	—	—	—	—	—	—	—	—	—
FRG	—	—	—	—	—	—	—	—	—
Gr.	—	—	—	—	—	—	—	—	—
Ice.	23.8	21.9	22.8	28.2	28.1	25.1	20.9	18.5	21.5
Ire.	13.3	4.0	1.7	1.1	0.4	—	—	—	—
It.	—	—	—	—	—	—	—	—	—
Lux.	—	—	—	—	—	—	—	—	—
Neth.	—	—	—	—	—	—	—	—	—
Norw.	8.0	9.1	9.3	9.4	10.2	11.0	8.6	6.7	6.6
Port.	—	—	—	—	—	—	—	—	—
Sp.	—	—	—	—	—	—	—	—	—
Sw.	12.4	10.7	11.1	13.6	16.5	22.5	21.1	15.5	10.6
Switz.	12.1	12.6	11.9	11.4	11.0	11.0	10.8	11.1	11.0
Turk.	—	—	—	—	—	—	—	—	—
UK	—	—	—	—	—	—	—	—	—
Can.	—	—	—	—	—	—	—	—	—
USA	—	—	—	—	—	—	—	—	—
Jap.	—	—	—	—	—	—	—	—	—
Austral.	11.1	9.1	8.6	8.7	9.2	9.7	10.7	9.6	11.5
NZ	—	—	—	—	—	—	—	—	—

Source: See Table 7.1.

TABLE 7.9. Electoral strength of left/socialist parties in national elections

	1945–9	1950–4	1955–9	1960–4	1965–9	1970–4	1975–9	1980–4	1985–9
Aus.	—	—	—	—	—	—	—	—	—
Belg.	—	—	—	—	—	—	—	—	—
Den.	—	—	—	6.0	9.5	9.1	11.7	14.1	16.9
Fin.	—	—	1.7	4.4	2.6	1.2	—	—	—
Fr.	—	—	0.7	2.4	3.1	3.3	3.3	1.3	—
FRG	—	—	—	—	—	—	—	—	—
Gr.	—	—	—	—	—	—	—	—	—
Ice.	—	6.0	3.5	—	—	—	—	—	—
Ire.	—	—	—	—	—	1.1	1.7	2.4	4.8
It.	—	—	—	—	4.4	2.6	1.5	1.5	1.7
Lux.	—	—	—	—	—	—	—	—	—
Neth.	—	—	0.9	3.0	2.9	1.5	0.9	2.2	1.6
Norw.	—	—	—	2.4	4.7	11.2	4.2	4.9	7.8
Port.	—	—	—	—	—	—	2.9	2.7	1.8
Sp.	—	—	—	—	—	—	1.0	—	—
Sw.	—	—	—	—	—	—	—	—	—
Switz.	—	—	—	—	—	0.3	1.7	2.7	1.8
Turk.	—	—	—	—	2.9	—	0.1	—	—
UK	—	—	—	—	—	—	—	—	—
Can.	—	—	—	—	—	—	—	—	—
USA	—	—	—	—	—	—	—	—	—
Jap.	—	—	—	—	—	—	—	—	—
Austral.	—	—	—	—	—	—	—	—	—
NZ	—	—	—	—	—	—	—	—	—

Source: See Table 7.1.

TABLE 7.10. Electoral strength of liberal parties in national elections

	1945–9	1950–4	1955–9	1960–4	1965–9	1970–4	1975–9	1980–4	1985–9
Aus.	11.7	10.6	7.1	7.0	5.4	5.5	5.8	5.0	9.7
Belg.	12.6	12.6	12.0	12.3	21.3	18.7	15.9	21.3	20.9
Den.	12.9	14.2	13.1	7.5	13.8	12.6	11.9	13.3	11.0
Fin.	4.6	7.1	6.2	6.4	6.5	5.6	4.0	—	0.9
Fr.	11.7	10.0	11.3	7.6	—	—	—	—	—
FRG	11.9	9.5	7.7	12.8	7.7	8.4	7.9	8.1	9.1
Gr.	39.7	43.7	35.8	44.0	—	20.4	12.0	1.5	0.1
Ice.	—	—	—	—	3.7	6.8	3.3	—	—
Ire.	47.4	47.8	50.7	45.3	46.7	46.2	50.6	45.9	52.9
It.	9.7	4.6	4.9	8.4	7.8	6.8	4.7	8.0	5.8
Lux.	15.5	10.8	18.5	10.6	16.6	22.2	21.3	18.7	16.2
Neth.	7.2	8.8	10.5	10.3	15.2	17.9	23.3	27.9	23.0
Norw.	13.5	10.0	9.6	8.8	9.9	6.9	4.6	4.4	3.4
Port.	—	—	—	—	—	—	27.7	22.6	52.7
Sp.	—	—	—	—	—	—	—	—	—
Sw.	22.7	24.4	21.0	17.5	15.3	12.8	10.9	5.9	13.2
Switz.	34.0	34.8	33.7	32.9	36.0	31.5	30.9	30.3	29.8
Turk.	—	—	—	—	2.8	1.1	0.4	—	0.8
UK	9.0	5.9	4.3	11.2	8.5	15.0	13.8	13.7	12.8
Can.	50.0	51.2	39.9	51.2	49.7	47.2	44.7	37.2	32.0
USA	52.1	44.5	42.1	55.5	42.7	37.5	50.3	41.4	45.6
Jap.	24.9	18.6	—	—	—	—	3.6	2.7	2.6
Austral.	0.6	—	—	—	—	0.8	5.0	5.7	6.0
NZ	—	11.1	7.2	8.3	11.8	6.7	11.8	14.2	5.7

Source: See Table 7.1.

TABLE 7.11a. Electoral strength of conservative parties in national elections (CDU/CSU, ÖVP, DC excluded)

	1945–9	1950–4	1955–9	1960–4	1965–9	1970–4	1975–9	1980–4	1985–9
Aus.	—	—	—	—	—	—	—	—	—
Belg.	—	—	—	—	—	—	—	—	—
Den.	15.3	18.2	18.91	21.9	20.6	13.0	8.8	19.0	20.1
Fin.	16.0	13.7	15.3	14.6	13.8	17.8	21.5	22.5	23.2
Fr.	13.5	35.7	31.0	45.8	44.7	37.0	36.2	32.2	42.7
FRG	4.0	3.3	3.4	2.8	0.1	—	—	—	—
Gr.	58.0	43.3	46.2	41.8	—	55.5	42.9	37.6	44.1
Ice.	39.5	40.4	41.5	41.4	38.6	40.5	34.1	38.7	27.2
Ire.	—	—	—	—	—	—	—	—	—
It.	2.8	6.8	4.8	1.7	1.3	—	—	—	—
Lux.	—	—	—	—	—	—	—	—	—
Neth.	—	—	—	—	—	—	—	—	—
Norw.	18.1	18.6	18.9	20.0	20.4	17.4	24.8	31.7	26.3
Port.	—	—	—	—	—	—	0.5	0.5	—
Sp.	—	—	—	—	—	—	42.4	35.3	35.1
Sw.	12.3	14.4	18.3	15.7	13.7	12.9	18.0	23.6	19.8
Switz.	—	—	—	—	—	—	—	—	—
Turk.	—	55.1	47.7	34.8	53.0	35.1	38.8	45.2	27.7
UK	39.6	45.7	49.6	43.4	41.9	40.1	43.9	42.4	42.3
Can.	28.6	31.0	46.3	35.1	31.9	35.2	35.9	41.3	42.9
USA	45.1	55.1	57.4	44.0	43.4	60.7	48.0	58.1	53.4
Jap.	31.7	47.8	60.5	55.8	48.2	46.9	43.2	46.9	49.4
Austral.	35.9	43.1	38.5	35.4	37.5	33.5	40.0	35.4	34.5
NZ	48.4	49.2	44.2	47.4	44.4	41.5	43.7	37.4	44.0

Source: See Table 7.1.

TABLE 7.11b. Electoral strength of conservative parties in national elections (CDU/CSU, ÖVP, DC included)

	1945–9	1950–4	1955–9	1960–4	1965–9	1970–4	1975–9	1980–4	1985–9
Aus.	46.9	41.3	45.1	45.4	48.3	43.9	42.4	43.0	41.3
Belg.	—	—	—	—	—	—	—	—	—
Den.	15.3	18.2	18.91	21.9	20.6	13.0	8.8	19.0	20.1
Fin.	16.0	13.7	15.3	14.6	13.8	17.8	21.5	22.5	23.2
Fr.	13.5	35.7	31.0	45.8	44.7	37.0	36.2	32.2	42.7
FRG	38.1	49.3	53.6	48.1	47.0	44.9	48.6	46.7	44.3
Gr.	58.0	43.3	46.2	41.8	—	55.5	42.9	37.6	44.1
Ice.	39.5	40.4	41.5	41.4	38.6	40.5	34.1	38.7	27.2
Ire.	—	—	—	—	—	—	—	—	—
It.	44.7	46.9	47.2	39.9	40.3	38.7	38.5	32.9	34.3
Lux.	—	—	—	—	—	—	—	—	—
Neth.	—	—	—	—	—	—	—	—	—
Norw.	18.1	18.6	18.9	20.0	20.4	17.4	24.8	31.7	26.3
Port.	—	—	—	—	—	—	0.5	0.5	—
Sp.	—	—	—	—	—	—	42.4	35.3	35.1
Sw.	12.3	14.4	18.3	15.7	13.7	12.9	18.0	23.6	19.8
Switz.	—	—	—	—	—	—	—	—	—
Turk.	—	55.1	47.7	34.8	53.0	35.1	38.8	45.2	27.7
UK	39.6	45.7	49.6	43.4	41.9	40.1	43.9	42.4	42.3
Can.	28.6	31.0	46.3	35.1	31.9	35.2	35.9	41.3	42.9
USA	45.1	55.1	57.4	44.0	43.4	60.7	48.0	58.1	53.4
Jap.	31.7	47.8	60.5	55.8	48.2	46.9	43.2	46.9	49.4
Austral.	35.9	43.1	38.5	35.4	37.5	33.5	40.0	35.4	34.5
NZ	48.4	49.2	44.2	47.4	44.4	41.5	43.7	37.4	44.0

Source: See Table 7.1.

TABLE 7.12. Electoral strength of protest parties in national elections

	1945–9	1950–4	1955–9	1960–4	1965–9	1970–4	1975–9	1980–4	1985–9
Aus.	—	—	—	—	3.3	0.3	—	—	—
Belg.	—	—	—	—	—	—	0.5	2.7	0.6
Den.	—	—	—	—	—	8.0	13.1	6.3	6.9
Fin.	—	—	—	2.2	1.0	9.9	5.1	9.8	7.5
Fr.	—	—	6.5	—	—	—	—	—	—
FRG	2.9	—	—	—	—	—	—	—	—
Gr.	—	—	—	—	—	—	—	—	—
Ice.	—	—	—	—	—	—	—	—	10.8
Ire.	—	—	—	—	—	—	—	—	—
It.	2.7	—	0.6	—	—	—	—	1.4	—
Lux.	—	2.8	—	6.0	0.4	—	4.5	—	7.3
Neth.	—	—	0.4	2.1	4.8	2.5	0.8	0.4	0.6
Norw.	—	—	—	—	—	5.0	1.9	4.5	8.4
Port.	—	—	—	—	—	—	—	—	—
Sp.	—	—	—	—	—	—	—	—	—
Sw.	—	—	—	—	—	—	—	—	—
Switz.	—	—	—	—	0.6	7.2	2.7	3.5	4.8
Turk.	—	—	—	—	—	—	—	—	—
UK	—	—	—	—	—	—	—	—	—
Can.	—	—	—	—	—	—	0.5	0.9	0.4
USA	—	—	—	—	13.5	1.4	0.2	—	0.5
Jap.	—	—	—	—	—	—	—	—	—
Austral.	—	—	—	—	—	—	—	—	—
NZ	—	—	—	—	—	—	—	—	—

Source: See Table 7.1.

TABLE 7.13. Electoral strength of ultra-right parties in national elections

	1945–9	1950–4	1955–9	1960–4	1965–9	1970–4	1975–9	1980–4	1985–9
Aus.	—	—	—	—	—	—	—	—	—
Belg.	—	—	—	—	—	—	—	—	—
Den.	—	—	—	—	—	—	—	—	—
Fin.	—	—	—	—	—	—	—	—	—
Fr.	1.8	—	1.0	0.8	0.1	0.5	0.5	0.3	9.9
FRG	—	1.1	—	—	3.2	0.6	0.3	0.2	0.6
Gr.	—	—	—	—	—	—	6.8	—	0.3
Ice.	—	—	—	—	—	—	—	—	—
Ire.	—	—	—	—	—	—	—	—	—
It.	1.0	5.8	4.8	5.1	4.4	8.7	5.7	6.8	5.9
Lux.	—	—	—	—	—	—	—	—	—
Neth.	—	—	—	—	—	—	—	—	—
Norw.	—	—	—	—	—	—	—	—	—
Port.	—	—	—	—	—	—	—	—	—
Sp.	—	—	—	—	—	—	1.3	—	—
Sw.	—	—	—	—	—	—	—	—	—
Switz.	—	—	—	—	—	—	—	—	—
Turk.	—	4.1	7.2	27.7	10.3	15.8	8.2	23.3	2.9
UK	—	—	—	—	—	0.2	0.6	0.1	—
Can.	—	—	—	—	—	—	—	—	—
USA	—	—	—	—	—	—	—	—	—
Jap.	—	—	—	—	—	—	—	—	—
Austral.	—	—	—	—	—	—	—	—	—
NZ	—	—	—	—	—	—	—	—	—

Source: See Table 7.1.

TABLE 7.14. Electoral strength of environmentalist/green parties in national elections

	1970–4	1975–9	1980–4	1985–9
Aus.	—	—	3.3	4.8
Belg.	—	0.6	4.4	6.7
Den.	—	—	—	1.3
Fin.	—	—	1.4	4.0
Fr.	—	2.2	1.4	0.8
FRG	—	—	3.6	8.3
Gr.	—	—	—	0.2
Ice.	—	—	5.5	11.7
Ire.	—	—	—	0.7
It.	—	2.3	2.2	5.1
Lux.	—	1.0	5.2	8.4
Neth.	—	—	—	0.2
Norw.	—	—	—	0.4
Port.	—	—	—	0.4
Sp.	—	—	—	0.7
Sw.	—	—	1.7	3.5
Switz.	—	—	2.9	7.8
Turk.	—	—	—	—
UK	—	—	0.2	0.3
Can.	—	—	—	0.4
USA	—	—	—	—
Jap.	—	—	—	—
Austral.	—	—	—	—
NZ	2.0	3.8	0.2	0.1

Source: See Table 7.1.

TABLE 7.15. Electoral strength of other parties in national elections

	1945–9	1950–4	1955–9	1960–4	1965–9	1970–4	1975–9	1980–4	1985–9
Aus.	0.4	0.4	0.1	0.5	0.0	0.1	0.0	0.3	0.3
Belg.	1.2	0.3	0.7	2.1	1.2	1.1	1.7	2.3	0.7
Den.	0.1	0.0	0.0	0.2	0.0	0.0	0.4	0.2	0.3
Fin.	0.4	0.1	0.5	0.7	0.0	0.1	0.5	0.0	0.6
Fr.	0.5	0.7	1.6	0.9	1.8	2.3	2.4	0.6	0.1
FRG	5.9	0.3	0.4	0.1	0.3	0.1	0.2	0.1	0.9
Gr.	2.3	3.0	0.8	0.5	—	1.1	0.9	0.5	0.8
Ice.	0.1	0.0	0.0	0.2	0.0	0.2	2.4	1.0	0.0
Ire.	8.3	7.7	6.6	5.4	2.7	3.9	5.6	3.8	4.0
It.	2.3	2.2	0.5	0.9	0.8	0.9	1.0	0.5	0.7
Lux.	0.9	0.0	0.7	0.0	0.0	1.1	0.4	0.2	4.1
Neth.	1.3	1.4	0.7	1.6	2.6	1.4	0.9	1.3	0.6
Norw.	0.1	0.0	0.2	0.2	0.1	0.9	0.9	0.9	0.8
Port.	—	—	—	—	—	—	1.6	1.2	1.0
Sp.	—	—	—	—	—	—	3.6	5.5	3.9
Sw.	0.1	0.1	0.1	0.1	0.0	0.5	0.6	0.3	0.6
Switz.	0.5	0.5	0.8	1.8	2.3	1.7	2.5	3.4	4.7
Turk.	—	—	—	0.8	4.4	2.8	2.5	1.1	0.4
UK	2.2	0.3	0.2	0.2	0.2	0.5	0.7	0.6	0.3
Can.	4.8	2.4	1.7	0.3	1.0	1.0	0.9	0.8	3.6
USA	0.0	0.1	0.5	0.4	0.2	0.2	1.4	0.5	0.5
Jap.	17.8	6.7	5.7	4.8	5.7	5.3	5.8	5.1	6.0
Austral.	2.3	1.4	1.1	0.7	2.1	0.8	1.5	1.7	2.0
NZ	0.4	0.3	0.2	0.6	0.9	1.5	0.8	1.1	1.4

Source: See Table 7.1.

PART II
Country Tables

1. Austria

State Structure: Federal.

Form of State: Republic.

Parliament (bicameral): Federal council (Bundesrat): 63 seats, 5/6 years; National council (Nationalrat): 183 seats, 4 years.

Electoral System: Proportional representation: Hare quota.

Main Language: German (98%).

Constitutional Development: Present constitution dates from 1920; amended 1929; also of importance is the state treaty concluded in 1955.

Heads of State:

K. Renner (SPÖ)	Dec. 1945–Dec. 1950
T. Körner (SPÖ)	May 1951–Jan. 1957
A. Schärf (SPÖ)	May 1957–Feb. 1965
F. Jonas (SPÖ)	May 1965–Apr. 1974
B. Kreisky (SPÖ) (interim)	Apr. 1974–July 1974
R. Kirchschläger (SPÖ)	July 1974–June 1986
K. Waldheim (ÖVP)	June 1986–

Capital City: Wien   (1985): 1,489,200.

Ministries:

Office of the Federal Chancellor

Agriculture and Forestry	Interior
Construction and Engineering	Justice
Education, the Arts, and Sport	National Defence
Family Affairs	Public Sector and Transport
Finance	Science and Research
Foreign Affairs	Social Affairs
Health and Environment	Trade, Commerce, and Industry

Governments:

	Prime Minister	Parties represented in government and distribution of ministers
Apr. 1945–Dec. 1945	K. Renner (SPÖ)	SPÖ (3), ÖVP (5), KPÖ (3)
Dec. 1945–Nov. 1947	L. Figl I (ÖVP)	SPÖ (5), ÖVP (6), KPÖ (1)
Nov. 1947–Nov. 1949	L. Figl II (ÖVP)	SPÖ (6), ÖVP (6)
Nov. 1949–Oct. 1952	L. Figl III (ÖVP)	SPÖ (5), ÖVP (5)
Oct. 1952–Apr. 1953	L. Figl IV (ÖVP)	SPÖ (4), ÖVP (5)
Apr. 1953–June 1956	J. Raab I (ÖVP)	SPÖ (4), ÖVP (5)
June 1956–July 1959	J. Raab II (ÖVP)	SPÖ (5), ÖVP (6)
July 1959–Nov. 1960	J. Raab III (ÖVP)	SPÖ (7), ÖVP (5)
Nov. 1960–Apr. 1961	J. Raab IV (ÖVP)	SPÖ (7), ÖVP (5)
Apr. 1961–Mar. 1963	A. Gorbach I (ÖVP)	SPÖ (7), ÖVP (5)
Mar. 1963–Apr. 1964	A. Gorbach II (ÖVP)	SPÖ (6), ÖVP (5)
Apr. 1964–Apr. 1966	J. Klaus I (ÖVP)	SPÖ (6), ÖVP (5)
Apr. 1966–Apr. 1970	J. Klaus II (ÖVP)	ÖVP (all)
Apr. 1970–Nov. 1971	B. Kreisky I (SPÖ)	SPÖ (all)
Nov. 1971–Oct. 1975	B. Kreisky II (SPÖ)	SPÖ (all)
Oct. 1975–May 1979	B. Kreisky III (SPÖ)	SPÖ (all)
May 1979–May 1983	B. Kreisky IV (SPÖ)	SPÖ (all)
May 1983–June 1986	F. Sinowatz (SPÖ)	SPÖ (12), FPÖ (3)
June 1986–Jan. 1987	F. Vranitzky I (SPÖ)	SPÖ (13), FPÖ (3)
Jan. 1987–	F. Vranitzky II (SPÖ)	SPÖ (8), ÖVP (8)

States (Länder):

	Population 1981	Capital
Burgenland	269 771	Eisenstadt
Kärnten	536 179	Klagenfurt
Niederösterreich	1 427 849	Sankt Pölten
Oberösterreich	1 269 540	Linz
Salzburg	442 301	Salzburg
Steiermark	1 186 525	Graz
Tirol	586 663	Innsbruck
Vorarlberg	305 164	Bregenz
Wien	1 531 346	(Wien)

Media:
(a) Major newspapers:

	Location	Orientation	Daily circulation c.1984
Neue Kronen-Zeitung	Wien	Independent	928 000
Kurier	Wien	Independent	495 000
Kleine Zeitung	Graz	Independent	157 000
Oberösterreichische Nachrichten	Linz	Independent	102 000
Tiroler-Tagezeitung	Innsbruck	Independent	95 000
Neue Zeit	Graz	SPÖ	80 000
Arbeiter-Zeitung	Wien	SPÖ	68 000
Die Presse	Wien	Independent	60 000
Salzburger Nachrichten	Salzburg	Independent	76 000
Süd-Ost-Tagepost	Graz	ÖVP	47 000
Volksstimme	Wien	KPÖ	46 000

(b) Radio and television: *Österreichischer Rundfunk* (ORF), founded in 1955; controls all radio and television: two national radio programmes, two national televison channels.

Economic Interest Organizations:
Employers' organization: Bundeskammer der gewerblichen Wirtschaft, founded 1946.

Central trade union peak organization: Österreichischer Gewerkschaftsbund (ÖGB), founded 1945; membership: 1,672,000 (1982).

Central Statistical Office:
Austrian Central Statistical Office
Hintere Zollamtsstr. 2B
A-1033 Vienna

Further Information:
Fischer, H. (1982) (ed.), *Das politische System Österreichs*. Wien: Europa Verlag.
Jelavich, B. (1987), *Modern Austria*. Cambridge: Cambridge U. P.
Nick, R., and Pelinka, A. (1983), *Bürgerkrieg—Sozialpartnerschaft: Das politische System Österreichs 1. und 2. Republik: Ein Vergleich*. Wien: Jugend und Volk.
Pelinka, A., and Plasser, F. (1987) (eds.), *Das Österreichischen Parteinsystem*. Wien: Böhlau.

Steiner, K. (1981) (ed.), *Modern Austria*. Palo Alto: SPOSS.
Sully, M. A. (1981), *Political Parties and Elections in Austria*. London: Hurst.
Sweeney, J., *et al.* (1988), *Austria: A Study in Modern Achievement*. Aldershot: Avebury.

2. Belgium

State Structure: Quasi-unitary (with regional and community levels of government).

Form of State: Constitutional monarchy.

Parliament (bicameral): Chamber of representatives (Chambre des représentants/Kamer van Volksvertegenwoordigers): 212 seats, 4 years; Senate (Sénat/Senaat): 183 seats, 4 years.

Electoral System: Proportional representation: Hare quota.

Main Languages: Dutch (59%); French (34%); German (1%).

Constitutional Development: Present constitution dates from 1830 but has been amended several times; changes were introduced in 1971, 1980, and 1988 with regard to the country's different cultural entities.

Heads of State:

Regency	Sept. 1944–July 1950
King Leopold III (abdication)	July 1950–July 1951
King Baudouin	July 1951–

Capital City: Bruxelles (1985): 976,536.

Ministries

Office of the Prime Minister

Brussels Region	Justice
Budget and Planning	National Defence
Civil Service	National Education (Flemish Sector)
Economic Affairs	
Employment and Labour	National Education (French Sector)
Finance and the Middle Classes	
Foreign Affairs	Posts and Telecommunications
Foreign Trade and Institutional Reforms (French Sector)	Public Works
	Scientific Policy
Interior	Social Affairs and Institutional Reform

Governments:

	Prime Minister	Parties represented in government and distribution of ministers
Feb. 1945–Aug. 1945	Van Acker I (BSP/PSB)	BSP/PSB (5), CVP/PSC (4), PVV/PLP(3), PCB/BCP (1)
Aug. 1945–Mar. 1946	Van Acker II (BSP/PSB)	BSP/PSB (5), PVV/PLP (5), UDB(1), PCB/BCP (1)
Mar. 1946–Apr. 1946	Spaak I (BSP/PSB)	BSP/PSB (all)
Apr. 1946–Aug. 1946	Van Acker III (BSP/PSB)	BSP/PSB (5), PVV/PLP (5), PCB/BCP (2)
Aug. 1946–Mar. 1947	Huysmans (BSP/PSB)	BSP/PSB (5), PVV/PLP(5), PCB/BCP (2)
Mar. 1947–Nov. 1948	Spaak II (BSP/PSB)	BSP/PSB (5), CVP/PSC (8)
Nov. 1948–Aug. 1949	Spaak III (BSP/PSB)	BSP/PSB (5), CVP/PSC (8)
Aug. 1949–June 1950	Eyskens I (CVP/PSC)	CVP/PSC (9), PVV/PLP (7)
June 1950–Aug. 1950	Duvieusart (CVP/PSC)	CVP/PSC (all)
Aug. 1950–Jan. 1952	Pholien (CVP/PSC)	CVP/PSC (all)
Jan. 1952–Apr. 1954	Van Houtte (CVP/PSC)	CVP/PSC (all)
Apr. 1954–June 1958	Van Acker (BSP/PSB)	BSP/PSB (9), PVV/PLP (7)
June 1958–Nov. 1958	Eyskens II (CVP/PSC)	CVP/PSC (all)
Nov. 1958–Sept. 1960	Eyskens III (CVP/PSC)	CVP/PSC (13), PVV/PLP (7)
Sept. 1960–Apr. 1961	Eyskens IV (CVP/PSC)	CVP/PSC (13), PVV/PLP (7)
Apr. 1961–July 1965	Lefèvre (CVP/PSC)	CVP/PSC (11), BSP/PSB (9)
July 1965–Mar. 1966	Harmel (CVP/PSC)	CVP/PSC (12),

	Prime Minister	Parties represented in government and distribution of ministers
Mar. 1966–June 1968	Vanden Boeynants (CVP/PSC)	BSP/PSB (8) CVP/PSC (12), PVV/PLP (7)
June 1968–Jan. 1972	Eyskens V (CVP/PSC)	CVP/PSC (15), BSP/PSB (13)
Jan. 1972–Jan. 1973	Eyskens VI (CVP)	PSB (5), BSP (4), CVP (5), PSC (4)
Jan.1973–Apr. 1974	Leburton (PSB)	PSB (4), BSP (4), CVP(5), PSC (4), PVV (2), PLP (2)
Apr. 1974–June 1974	Tindemans I (CVP)	CVP (7), PSC (6), PVV(3),PLP (3)
June 1974–June 1977	Tindemans II (CVP)	CVP (7), PSC (6), PVV (3),PLP (3), RW (1)
June 1977–Oct. 1978	Tindemans III (CVP)	CVP (5), PSC (4), PSB(5),BSP (4), FDF (2), VU (2)
Oct. 1978–Apr. 1979	Vanden Boeynants II (CVP) (PSC)	CVP (5), PSC (4), PSB (5), BSP (4), FDF (2), VU (2)
Apr. 1979–Jan. 1980	Martens I (CVP)	CVP (7), PSC (6), PSB(5),BSP (5), FDF (1)
Jan. 1980–May 1980	Martens II (CVP)	CVP (7), PSC (6), PSB(6), BSP (5)
May 1980–Oct. 1980	Martens III (CVP)	CVP (6), PSC (5), PSB(5), BSP (4), PVV (2), PLP (2)
Oct. 1980–Apr. 1981	Martens IV (CVP)	CVP (7), PSC (5), PSB (7), BSP (5)
Apr. 1981–Sept. 1981	Eyskens I (CVP)	CVP (7), PSC (5), PSB(7), BSP (5)
Sept. 1981–Dec. 1981	Eyskens II (CVP)	caretaker
Dec. 1981–Nov. 1985	Martens V (CVP)	CVP (8), PSC (5),

	Prime Minister	Parties represented in government and distribution of ministers
Nov. 1985–May 1988	Martens VI (CVP)	PVV(6), PRL (6) CVP (10), PSC (6),
May 1988–	Martens VII (CVP)	PVV (6), PRL (6) CVP (5), PSC (3), PS (6), SP (3), VU (2)

Provinces:

	Population 1983	Capital
Antwerp	1 578 869	Antwerp
Brabant	2 217 442	Brussels
Flanders (East)	1 331 193	Ghent
Flanders (West)	1 086 574	Bruges
Hainaut	1 285 936	Mons
Liège	992 061	Liège
Limburg	726 884	Hasselt
Luxembourg	223 813	Arlon
Namur	410 251	Namur

Media:
(*a*) Major newspapers:

	Location	Orientation	Daily circulation in 1981
De Standaard	Brussels	CVP	338 000
Het Laatste Nieuws	Brussels	Independent	314 000
Le Soir	Brussels	Independent	219 000
Het Volk	Ghent	CVP	205 000
De Gazet van Antwerpen	Antwerp	CVP	196 000
La Meuse	Liège	Independent	146 000
La Lanterne	Brussels	PSB	134 000
La Libre Belgique	Brussels	PSC	125 000
La Derniere Heure	Brussels	PRL	125 000
La Nouvelle Gazette	Charleroi	PRL	75 000

(b) Radio and television: Public: *Radio-Télévision Belge de la Communauté Culturelle Française* (French); *Belgische Radio en Televisie* (Dutch). Private: *Radio Télé Luxembourg* (RTL) (French); *VTM* (Dutch).

Economic Interest Organizations:
Employers' organization: Fédération des Entreprises de Belgique.

Central trade union peak organizations: Fédération Génerale du Travail de Belgique/Algemeen Belgisch Vakverbond (FGTB), founded 1898, reorganized 1937; membership: 1,115,290 (1984); Confédération des Syndicats Chrétiens (CSC), founded 1909; membership 1,336,000 (1984).

Central Statistical Office:
Institut National de Statistique
44 rue du Louvain,
B-Brussels 1000

Further Information:
Albareuol *et al.* (1987), 'Belgique', in *Cahiers du CACEF*, 130. CACEF.
Delruelle *et al.* (1983), *Regioscope I & IV: 1979-1982*. Bruxelles: CRISP.
Fitzmaurice, J. (1983), *The Politics of Belgium: Crisis and Compromise in a Plural Society*. London: Hurst.
Mabille, X. (1985), *Histoire politique de la Belgique: Facteurs et acteurs de changement*. Bruxelles: CRISP.
McRae, K. D. (1983), *Conflict and Compromise in Multilingual Societies: Belgium*. Waterloo, Ont.: Wilfred Laurier U. P.

3. Denmark

State Structure: Unitary.

Form of State: Constitutional monarchy.

Parliament (unicameral): Folketing: 179 seats, 4 years.

Electoral System: Proportional representation: modified Sainte Lague.

Main Language: Danish (98%).

Constitutional Development: The constitutional charter (Grundloven) was adopted in 1953. The previous constitution dates from 1849 (changed 1866 and 1915).

Heads of State:

King Christian X	May 1912–Apr. 1947
King Frederik IX	Apr. 1947–Jan. 1972
Queen Margrethe II	Jan. 1972–

Capital City: København (1985): 1,358,540.

Ministries

Office of the Prime Minister

Agriculture	Foreign Affairs
Cultural Affairs	Greenland
Defence	Housing
Ecclesiastical Affairs	Industry
Economic Affairs	Inland Revenue
Education	Interior
Energy	Justice
Environment	Labour
Finance	Social Affairs
Fisheries	Transport and Public Works

Governments:

	Prime Minister	Parties represented in government and distribution of ministers
May 1945–Nov. 1945	Buhl (SD)	SD (4), V (2), RV (1), KF (2), DKP (2), others (7)
Nov. 1945–Nov. 1947	Kristensen (V)	V (all)
Nov. 1947–Sept. 1950	Hedtoft I (SD)	SD (all)
Sept. 1950–Oct. 1950	Hedtoft II (SD)	SD (all)
Oct. 1950–Sept. 1953	Eriksen (V)	V (7), KF (6)
Sept. 1953–Feb. 1955	Hedtoft III (SD)	SD (all)
Feb. 1955–May 1957	Hansen I (SD)	SD (all)
May 1957–Feb. 1960	Hansen II (SD)	SD (10), RV (4), RFB (3)
Feb. 1960–Nov. 1960	Kampmann I (SD)	SD (10), RV (4), RFB (3)
Nov. 1960–Sept. 1962	Kampmann II (SD)	SD (11), RV (5)
Sept. 1962–Sept. 1964	Krag I (SD)	SD (12), RV (5)
Sept. 1964–Nov. 1966	Krag II (SD)	SD (all)
Nov. 1966–Feb. 1968	Krag III (SD)	SD (all)
Feb. 1968–Oct. 1971	Baunsgaard (RV)	RV (8), V (8), KF (6)
Oct. 1971–Oct. 1972	Krag IV (SD)	SD (all)
Oct. 1972–Dec. 1973	Anker Jörgensen I (SD)	SD (all)
Dec. 1973–Feb. 1975	Hartling (V)	V (all)
Feb. 1975–Feb. 1977	Anker Jörgensen II (SD)	SD (all)
Feb. 1977–Aug. 1978	Anker Jörgensen III (SD)	SD (all)
Aug. 1978–Oct. 1979	Anker Jörgensen IV (SD)	SD (14), V (7)
Oct. 1979–Dec. 1981	Anker Jörgensen V (SD)	SD (all)
Dec. 1981–Sept. 1982	Anker Jörgensen VI (SD)	SD (all)
Sept. 1982–Jan. 1984	Schlüter I (KF)	KF (8), V (8), KRF (1), CD (4)
Jan. 1984–Sept. 1987	Schlüter II (KF)	KF (8), V (8), KRF (1), CD (4)
Sept. 1987–June 1988	Schlüter III (KF)	KF (9), V (8), KRF (2), CD (3)
June 1988–	Schlüter IV (KF)	KF (9), V (8), RV (4)

Provinces (Amt):

	Population 1983	Capital
Københavns kommune	486 593	—
Frediksborg	88 409	—
Københavns amtskommune	619 687	—
Fredriksborg amtskommune	331 349	Hillerød
Roskilde amtskommune	205 414	Roskilde
Vestsjælland amtskommune	277 914	Sorø
Storstrøm amtskommune	258 670	Nykøbing
Bornholm amtskommune	47 313	Rønne
Fyn amtskommune	453 773	Odense
Sønderjylland amtskommune	249 970	Åbenrå
Ribe amtskommune	214 700	Ribe
Vejle amtskommune	327 102	Vejle
Ringkøbing amtskommune	264 103	Ringkøbing
Århus amtskommune	578 149	Århus
Viborg amtskommune	230 909	Viborg
Nordjylland amtskommune	482 409	Ålborg

Media:
(a) Major newspapers:

	Location	Orientation	Daily circulation
Ekstrabladet	København	Independent V	249 000
B.T.	København	Independent KF	214 000
Politiken	København	Independent RV	141 000
Berlingske Tidende	København	Independent KF	117 000
Jyllands-Posten	Århus	Independent	101 000
Ålborg Stiftstidende	Ålborg	Independent	73 000
Århus Stiftstidende	Århus	KF	69 000
Fyens Stiftstidende	Odense	KF	68 000
Aktuelt	København	SD	58 000
Land og Folk	København	DKP	10 000

(b) Radio and television: *Radio Danmark, TV2*.

Economic Interest Organizations:
Employers' organization: Dansk arbejdsgiverforening.
  Central trade union peak association: Landsorganisationen i Danmark (LO), founded 1898; membership: 1,411,000 (1984); Funktionærernes

og Tjenestemændens Fællesråd (FTF), founded 1952; membership: 350,000 (1984).

Central Statistical Office:
Danmarks Statistik
Sejrøgade 11
DK-2100 København Ø

Further Information:
Daalder, H. (1987) (ed.), *Party Systems in Denmark, Austria, Switzerland, The Netherlands and Belgium*. London: Pinter.
Elder, N., *et al.* (1988), T*he Consensual Democracies: The Government and Politics of the Scandinavian States*. Rev. edn. Oxford: Blackwell.
Eliassen, K., and Pedersen, M. N. (1985), *Nordiske politiske fakta*. Oslo: Tiden.
Fitzmaurice, J. (1981), *Politics in Denmark*. London: Hurst.

4. Finland

State Structure: Unitary.
Form of State: Republic.
Parliament (unicameral): Eduskunta: 200 seats, 4 years.
Electoral System: Proportional representation: d'Hondt divisor.
Main Languages: Finnish (94%); Swedish (6%).
Constitutional Development: Present constitution dates from 1919.
Heads of State:

J. Paasikivi (KOK)   Mar. 1945–Feb. 1956
U. Kekkonen (KESK) Mar. 1956–Jan. 1982
M. Koivisto (SDP)   Jan. 1982–

Capital City: Helsinki (1985): 484,263.

Ministries

Office of the Prime Minister

Agriculture and Forestry	Interior
Defence	Justice
Education	Labour
Environment	Social Affairs and Health
Finance	Trade and Industry
Foreign Affairs	Transport and Communications

Governments:

	Prime Minister	Parties represented in government and distribution of ministers
Apr. 1945–Mar. 1946	Paasikivi III (Indep.)	SKDL (5), SDP (4), KESK (4), SFP (1), LKP (1)

Finland

	Prime Minister	Parties represented in government and distribution of ministers
Mar. 1946–July 1948	Pekkala (SKDL)	SKDL (6), SDP (5), KESK (5), SFP (1)
July 1948–Mar. 1950	Fagerholm I (SDP)	SDP (all)
Mar. 1950–Jan. 1951	Kekkonen I (KESK)	KESK (10), LKP (2), SFP (3)
Jan. 1951–Sept. 1951	Kekkonen II (KESK)	SDP (7), KESK (7), LKP (1), SFP (2)
Sept. 1951–July 1953	Kekkonen III (KESK)	SDP (7), KESK (7), LKP (1), SFP (2)
July 1953–Nov. 1953	Kekkonen IV (KESK)	KESK (8), SFP (3)
Nov. 1953–May 1954	Tuomioja (Indep.)	LKP (3), SFP (2), KOK (4)
May 1954–Oct. 1954	Törngren (SFP)	SDP (6), KESK (6), SFP (1)
Oct. 1954–Mar. 1956	Kekkonen V (KESK)	SDP (7), KESK (6)
Mar. 1956–May 1957	Fagerholm II (SDP)	SDP (6), KESK (6), LKP (1), SFP (1)
May 1957–Nov. 1957	Sukselainen (KESK)	KESK (6), LKP (3), SFP (3)
Nov. 1957–Apr. 1958	von Fieandt (Indep.)	KESK (4)
Apr. 1958–Aug. 1958	Kuuskoski (Indep.)	KESK (5), TPSL (4), LKP (1)
Aug. 1958–Jan. 1959	Fagerholm III (SDP)	SDP (5), KESK (5), LKP (1), SFP (1), KOK (3)
Jan. 1959–July 1961	Sukselainen II (KESK)	KESK (14), SFP (1)
July 1961–Apr. 1962	Miettunen I (KESK)	KESK (14)

	Prime Minister	Parties represented in government and distribution of ministers
Apr. 1962–Dec. 1963	Karjalainen I (KESK)	KESK (5), LKP (2), SFP (2), KOK (3)
Dec. 1963–Sept. 1964	Lehto (Indep.)	SDP (1)
Sept. 1964–May 1966	Virolainen (KESK)	KESK (7), SFP (2), LKP (3), KOK (3)
May 1966–Mar. 1968	Paasio I (SDP)	KESK (5), SDP (6), TPSL (1), SKDL (3)
Mar. 1968–May 1970	Koivisto I (SDP)	SDP (6), KESK (5), TPSL (1), SFP (1), SKDL (3)
May 1970–July 1970	Aura I (LKP)	SDP (4), KESK (3), LKP (1), SFP (1), KOK (1)
July 1970–Oct. 1971	Karjalainen II (KESK)	KESK (5), SDP (5–8), LKP(2), SFP (2), SKDL (3–0)
Oct. 1971–Feb. 1972	Aura II (LKP)	SDP (3), KESK (4), LKP (1), SFP (1), KOK (1)
Feb. 1972–Sept. 1972	Paasio II (SDP)	SDP (all)
Sept. 1972–June 1975	Sorsa I (SDP)	SDP (7), KESK (5), LKP(1), SFP (2)
June 1975–Nov. 1975	Liinamaa (SDP)	SDP (5), KESK (3), LKP (1), KOK (1), SFP (1)

Finland

	Prime Minister	Parties represented in government and distribution of ministers
Nov. 1975–Sept. 1976	Miettunen II (KESK)	SDP (5), KESK (4), LKP (1), SFP (2), SKDL (4)
Sept. 1976–May 1977	Miettunen III (KESK)	KESK (9), LKP (3), SFP (3)
May 1977–Mar. 1978	Sorsa II (SDP)	SDP (4), KESK (5), LKP (1), SFP (1), SKDL (3)
Mar. 1978–May 1979	Sorsa III (SDP)	SDP (4), KESK (5), LKP (2), SKDL (3)
May 1979–Feb. 1982	Koivisto II (SDP)	SDP (5), KESK (6), SFP (2), SKDL (3)
Feb. 1982–Dec. 1982	Sorsa IV (SDP)	SDP (5), KESK (6), SFP (2), SKDL (3)
Dec. 1982–May 1983	Sorsa V (SDP)	SDP (8), KESK (6), SFP (2)
May 1983–Apr. 1987	Sorsa VI (SDP)	SDP (8), KESK (5), SFP (2), SMP (2)
Apr. 1987–	Holkeri (KOK)	KOK (7), SDP (8), SFP (2), SMP (1)

Provinces (Läänit):

	Population 1984	Capital
Uusimaa	1 175 373	Helsinki
Turku ja Pori	712 439	Turku
Ahvenanmaa	23 595	Maarianhamina
Häme	675 127	Hämeenlinna

	Population 1984	Capital
Kymi	341 709	Kouvala
Mikkeli	209 256	Mikkeli
Kuopio	255 740	Kuopio
Pohjois-Karjala	177 633	Joensuu
Vaasa	443 743	Vaasa
Keski-Suomi	247 351	Jyväskylä
Oulu	430 902	Oulu
Lappi	200 879	Rovaniemi

Media:
(a) Major newspapers

	Location	Orientation	Daily circulation
Helsingin Sanomat	Helsinki	Independent	457 068
Aamulehti	Tampere	KOK	142 920
Turun Sanomat	Turku	Independent	133 489
Maaseudun Tulevaisuus	Helsinki	Independent	129 173
Ilta-Sanomat	Helsinki	Independent	195 280
Uusi Suomi	Helsinki	Independent	90 322
Savon Sanomat	Kuopio	KESK	87 819
Keskisuomalainen	Jyväskylä	KESK	79 362
Hufvudstadsbladet	Helsinki	Independent	66 562
Kansan Uutiset	Helsinki	SKDL	43 454
Suomen Sosiaalidemokraatti	Helsinki	SDP	32 669

(b) Radio and television: *Öy Yleisradio AB* (YLE) (Finnish Broadcasting Company). The state operates three television channels, and broadcasting time is leased from them by commercial companies, e.g. MTV. There are also commercial local radios and local cable television companies have their own programmes.

Economic Interest Organizations:
Employers' organization: Suomen Työnantajain Keskusliitto (Finnish Employers' Confederation).

Central trade union peak association: Suomen Ammattiliittojen Keskusjärjestö (Central Organization of Finnish Trade Unions) (SAK), first founded 1907, reunited 1969; membership: 1,040,000 (1984); Toimihenkilö-ja Virkamiesjärjestöjen Keskusliitto (Confederation of Salaried Employees) (TVK), founded 1927; membership: 359,000 (1984).

Central Statistical Office:
Central Statistical Office of Finland
POB 504
SF-00101 Helsinki

Further Information:
Arter, D. (1987), *Politics and Policy-Making in Finland*. Brighton: Wheatsheaf.
Kirby, D. G. (1979), *Finland in the Twentieth Century*. London: Hurst.
Lindström, U., and Karvonen, L. (1987) (eds.), *Finland: En politisk loggbok*. Stockholm: Almqvist & Wiksell International.

5. France

State Structure: Unitary.

Form of State: Republic.

Parliament (bicameral): National assembly (Assemblée nationale): 577 seats, 5 years; Senate (Sénat): 319 seats, 9 years.

Electoral System: First round: majority formula; second round: plurality formula.

Main Languages: French (93%); Arabic (3%).

Constitutional Development: The present constitution of the Fifth French Republic dates from 1958, with modification in 1962 (election of the President).

Heads of State:

A. Auriol (SFIO)	Jan. 1947–Dec. 1953
R. Coty (Moderate Indep.)	Dec. 1953–Oct. 1958
C. de Gaulle (RPR)	Jan. 1959–Apr. 1969
A. Poher (Interim)	Apr. 1969–June 1969
G. Pompidou (UDR)	June 1969–Apr. 1974
A. Poher (Interim)	Apr. 1974–May 1974
V. Giscard d'Estaing (PR)	May 1974–May 1981
F. Mitterrand (PS)	May 1981–

Capital City: Paris (1982): 2,188,960.

Ministries

Office of the President
Office of the Prime Minister

Agriculture
Defence
Economy, Finance, and the Budget
Environment
External Relations
Industrial Redeployment and Foreign Trade
Interior and Decentralization
Justice
Labour, Employment, and
Vocational Training
National Education
Planning and Regional Development
Research and Technology
Social Affairs and National Solidarity
Town Planning, Housing, and Transport
Trade, Artisan Industries, and Tourism

Governments:

	Prime Minister	Parties represented in government and distribution of ministers
Nov. 1944–Nov. 1945	de Gaulle I	PCF, SFIO, MRP, Rad. Soc.
Nov. 1945–Jan. 1946	de Gaulle II	PCF, MRP, SFIO, UDSR, Rad. Soc.
Jan. 1946–June 1946	Gouin (SFIO)	SFIO, MRP, PCF
June 1946–Dec. 1946	Bidault I (MRP)	SFIO, MRP, PCF, UDSR
Dec. 1946–Jan. 1947	Blum (SFIO)	SFIO (all)
Jan. 1947–Nov. 1947	Ramadier (SFIO)	SFIO, MRP, PCF, UDSR, Soc. Ind., Rad. Soc.
Nov. 1947–July 1948	Schuman I (MRP)	MRP, SFIO, Rad. Soc., UDSR, Soc. Ind.
July 1948–Aug. 1948	Marie (Rad. Soc.)	SFIO, MRP, UDSR, PRL, Rad. Soc., Soc. Ind.
Aug. 1948–Sept. 1948	Schuman II (MRP)	MRP, Rad. Soc., Soc. Ind., UDSR
Sept. 1948–Oct. 1949	Queuille I (Rad. Soc.)	SFIO, MRP, Rad. Soc., UDSR, Indep., PRL
Oct. 1949–Feb. 1950	Bidault II (MRP)	MRP (6), Rad. Soc. (2), SFIO (6), UDSR (2), Indep. (1)
Feb. 1950–July 1950	Bidault III (MRP)	MRP (10), Rad. Soc. (3), UDSR (2), Indep. (2)
July 1950–July 1950	Queuille II (Rad. Soc.)	MRP (9), Rad. Soc. (6), Indep. (2), UDSR (2)
July 1950–Mar. 1951	Pleven I (UDSR)	MRP (9), UDSR (3),

	Prime Minister	Parties represented in government and distribution of ministers
Mar. 1951–Aug. 1951	Queuille III (Rad. Soc.)	SFIO (9), Indep. (3), Rad. Soc. (8) MRP (7), UDSR (3),
Aug. 1951–Jan. 1952	Pleven II (UDSR)	SFIO (5), Rad. Soc. (4), Indep. (2) MRP (7), Indep. (6), Rad. Soc. (5), UDSR (2), Paysan (2), RGR (1)
Jan. 1952–Mar. 1952	Faure I (Rad. Soc.)	Rad. Soc. (8), MRP (8), UDSR (3), Indep. (5), Paysan (2)
Mar. 1952–Jan. 1953	Pinay (Indep.)	Rad. Soc. (5), Indep. (4), MRP (4), Paysan (2), UDSR (2)
Jan. 1953–June 1953	Mayer (Rad. Soc.)	Rad. Soc. (7), MRP (6), Indep. (4), Paysan (3), UDSR (2), ARS (1)
June 1953–June 1954	Laniel (Indep.)	Indep. (5), MRP (5), Rad. Soc. (4), Paysan (2), UDSR (2), ARS (1), URAS (3)
June 1954–Feb. 1955	Mendes-France (Rad. Soc.)	Rad. Soc. (4), Gaull. (4), Indep. (3), UDSR (2),

France 173

	Prime Minister	Parties represented in government and distribution of ministers
Feb. 1955–Jan. 1956	Faure II (Rad. Soc.)	MRP (1), RGR (1) Rad. Soc. (5), Gaull. (4), Indep. (2), MRP (4), Paysan (2), UDSR (1)
Jan. 1956–June 1957	Mollet (SFIO)	SFIO (6), Rad. Soc. (4), UDSR (1)
June 1957–Nov. 1957	Bourges-Maunoury (Rad. Soc.)	Rad. Soc. (6), SFIO (5), UDSR (1), RGR (1)
Nov. 1957–May 1958	Gaillard (Rad. Soc.)	Rad. Soc. (3), MRP (3), SFIO (4), Indep. (3), UDSR (1), RGR (1)
May 1958–June 1958	Pfimlin (MRP)	MRP (4), UDSR (2), RGR (2), Indep. (4), Rad. Soc. (3)
June 1958–Jan. 1959	de Gaulle III (UNR)	SFIO (3), MRP (3), UDSR (1), Indep. (3), Gaull. (3), Rad. Soc. (2)
Jan. 1959–Apr. 1962	Debré (UNR)	UNR (6), MRP (3), Indep. (3), Rad. Soc. (1)
Apr. 1962–Dec. 1962	Pompidou I (UNR)	UNR (7), MRP (4),

	Prime Minister	Parties represented in government and distribution of ministers
Dec. 1962–Jan. 1966	Pompidou II (UNR–UDT)	Indep. (1) UNR–UDT (12), Rep. Ind. (2)
Jan. 1966–Apr. 1967	Pompidou III (UNR–UDT)	UNR–UDT (8), Rep. Ind. (1), GD (2)
Apr. 1967–May 1968	Pompidou IV (UNR–UDT)	UNR–UDT (16), Rep. Ind. (2)
May 1968–July 1968	Pompidou V (UNR–UDT)	UNR–UDT (14), Rep. Ind. (3)
July 1968–June 1969	Couve de Murville (UDR)	UDR (15), Rep. Ind. (3)
June 1969–July 1972	Chaban-Delmas (UDR)	UDR (12), Rep. Ind. (4), PDM (3)
July 1972–Apr. 1973	Messmer I (UDR)	UDR (14), Rep. Ind. (3), CDP (3)
Apr. 1973–Mar. 1974	Messmer II (UDR)	UDR (13), Rep. Ind. (4), CDP (2)
Mar. 1974–May 1974	Messmer III (UDR)	UDR (15), Rep. Ind. (3), CDP (1)
May 1974–Aug. 1976	Chirac I (UDR)	UDR (15), Rep. Ind. (4), MRG (4)
Aug. 1976–Mar. 1978	Barre I (UDR)	UDR (5), Rep. Ind. (4), CDS (1), Rad. Soc. (3)
Mar. 1978–May 1981	Barre II (RPR)	RPR (4), Rep. Ind. (3), CDS (2), Rad. Soc. (1)
May 1981–June 1981	Mauroy I (PS)	PS (29), MRG (2)
June 1981–Mar. 1983	Mauroy II (PS)	PS (29), PCF (4), MRG (1)
Mar. 1983–July 1984	Mauroy III (PS)	PS (36), PCF (4), MRG (1), PSU (1)
July 1984–Mar. 1986	Fabius (PS)	PS (39), MRG (1), PSU (1)

	Prime Minister	Parties represented in government and distribution of ministers
Mar. 1986–May 1988	Chirac II (RPR)	RPR (20), PR (7), CDS (7)
May 1988–June 1988	Rocard I (PS)	PS (24), MRG (2), UDF (2), Indep. (13)
June 1988–	Rocard II (PS)	

Regions (Régions):

	Population 1982	Capital
Île-de-France	10 064 840	Paris
Champagne-Ardennes	1 344 820	Reims
Picardie	1 740 460	Amiens
Haute-Normandie	1 659 520	Rouen
Centre	2 265 340	Orléans
Basse-Normandie	1 350 480	Caen
Bourgogne	1 592 300	Dijon
Nord-Pas-de-Calais	3 919 240	Lille
Lorraine	2 334 740	Nancy
Alsace	1 553 740	Strasbourg
Franche-Comté	1 078 700	Besançon
Pays de la Loire	2 937 980	Nantes
Bretagne	2 703 440	Rennes
Poitou-Charentes	1 567 600	Poitiers
Aquitaine	2 655 800	Bordeaux
Limousin	736 340	Limoges
Rhône-Alpes	5 022 800	Lyon
Auvergne	1 329 180	Clermont-Ferrand
Languedoc-Roussillon	1 929 520	Montpellier
Provence-Alpes-Cote d'Azur	3 942 980	Marseille
Corse	234 640	Ajaccio

Media:
(*a*) Major newspapers:

	Location	Orientation	Daily circulation
Ouest-France	Rennes	Mass	1 676 000
Le Monde	Paris	Independent	425 000
France-soir	Paris	Mass	403 000
Sud-ouest	Bordeaux	Independent	355 000
La Montagne	Clermont-Ferrand	Independent	351 000
Le Parisien Libéré	Paris	Right	333 000
Le Progrès	Lyon	Conservative	331 000
Le Dauphiné Libéré	Grenoble	Conservative	312 000
Le Figaro	Paris	Right	300 000
L'Humanité	Paris	Communist	138 000
Le Matin de Paris	Paris	Socialist	136 000
La Voix du Nord	Lille	Independent	376 000
La Dépêche du Midi	Toulouse	Rad. Soc.	255 000
Le Provençal	Marseille	Socialist	172 000
Libération	Paris	Independent left	70 000

(*b*) Radio and television: *Commission Nationale de la Communication et des Libertés (CNCL)* supervises all French broadcasting, allocates concessions for privatized channels, distributes cable networks and frequencies. In 1987 there were two state-run channels (*A2* and *FR3*) and four private channels (*Canal Plus*, *La 5*, *M6*, and *TF1*).

Economic Interest Organizations:
Employers' organization: Conseil National du Patronat Français (CNPF).

Central trade union peak associations: Confédération Générale du Travail (CGT), founded 1895; membership: 1,600,000 (1984); Force Ouvrière (FO), founded 1947; membership: 1,000,000 (1984); Confédération Française Démocratique du Travail (CFDT), founded 1964; membership: 900,000 (1984); Confédération Générale des Cadres (CGC), founded 1944; membership: 300,000 (1984); Fédération de l'Éducation Nationale (FEN), founded 1948; membership: 500,000 (1984).

Central Statistical Office:
Institut National de la Statistique et des Études Économiques

18, boulevard Adolphe-Pinard
F-75675 PARIS CEDEX 14

Further Information:

Ardagh, J. (1987), *France Today*. London: Secker & Warburg.

Hayward, J. (1983), *Governing France: The One and Indivisible French Republic*. London: Weidenfeld & Nicolson.

Larkin, M. (1988), *France since the Popular Front: Government and People 1936-86*. Oxford: Oxford U. P.

Ross, G. *et al.* (1987) (eds.), *The Mitterrand Experiment*. Oxford: Polity Press.

Wright, V. (1983), *The Government and Politics of France*. London: Hutchinson.

6. Federal Republic Of Germany[1]

State Structure: Federal.

Form of State: Republic.

Parliament (bicameral): Federal assembly (Bundestag): 520 seats, 4 years; Federal council (Bundesrat): 45 seats, various no. of years.

Electoral System: First vote (constituency): plurality formula; second vote (national): proportional vote: d'Hondt.

Main Languages: German (94%); Turkish (2%).

Constitutional Development: Present constitution dates from 1949.

Heads of State:

T. Heuss (FDP)	Sept. 1949–July 1959
H. Lubke (CDU)	July 1959–June 1969
G. Heinemann (SPD)	July 1969–June 1974
W. Scheel (FDP)	July 1974–June 1979
K. Carstens (CDU)	July 1979–June 1984
R. von Weizäcker (CDU)	July 1984–

Capital City: Bonn (1985): 292,600.

Ministries

Office of the Federal President
Office of the Federal Chancellor

Defence	Justice
Economic Co-operation	Labour and Social Affairs
Economics	Posts and Telecommunications
Education and Science	Regional Planning, Construction
Finance	and Urban Development
Food, Agriculture, and Forestry	Research and Technology
Foreign Affairs	Transport
Interior	Youth, Family, and Health Affairs
Intra-German Relations	

---

1. These data refer to the Federal Republic before reunification.

Governments:

	Prime Minister	Parties represented in government and distribution of ministers
Sept. 1949–Sept. 1953	Adenauer I (CDU)	CDU/CSU (9), FDP (3), DP (2)
Oct. 1953–Oct. 1957	Adenauer II (CDU)	CDU/CSU (10), FDP (4), DP (2), BHV (2)
Oct. 1957–Nov. 1961	Adenauer III (CDU)	CDU/CSU (16), DP (2)
Nov. 1961–Dec. 1962	Adenauer IV (CDU)	CDU/CSU (16), FDP (5)
Dec. 1962–Oct. 1963	Adenauer V (CDU)	CDU/CSU (16), FDP (5)
Oct. 1963–Oct. 1965	Erhard I (CDU)	CDU/CSU (16), FDP (5)
Oct. 1965–Dec. 1966	Erhard II (CDU)	CDU/CSU (18), FDP (4)
Dec. 1966–Oct. 1969	Kiesinger (CDU)	CDU/CSU (11), SPD (9)
Oct. 1969–Dec. 1972	Brandt I (SPD)	SPD (12), FDP (3)
Dec. 1972–May 1974	Brandt II (SPD)	SPD (13), FDP (5)
May 1974–Dec. 1976	Schmidt I (SPD)	SPD (12), FDP (4)
Dec. 1976–Nov. 1980	Schmidt II (SPD)	SPD (12), FDP (4)
Nov. 1980–Oct. 1982	Schmidt III (SPD)	SPD (13), FDP (4)
Oct. 1982–Mar. 1983	Kohl I (CDU)	CDU/CSU (14), FDP (3)
Mar. 1983–Mar. 1987	Kohl II (CDU)	CDU/CSU (14), FDP (3)
Mar. 1987–	Kohl III (CDU)	CDU/CSU (15), FDP (4)

States (Länder):

	Population 1983	Capital
Schleswig-Holstein	2 615 100	Kiel
Hamburg	1 600 300	Hamburg
Niedersachsen	7 229 900	Hannover
Bremen	671 600	Bremen
Nordrhein-Westfalen	16 775 900	Düsseldorf
Hessen	5 548 700	Wiesbaden
Rheinland-Pfalz	3 627 800	Mainz
Baden-Württemberg	9 242 800	Stuttgart
Bayern	10 965 800	München
Saarland	1 051 600	Saarbrucken
West Berlin	1 851 800	West Berlin

Media:
(*a*) Major newspapers:

	Location	Orientation	Daily circulation
Bild	Hamburg	Left of centre	5 400 000
Berliner Morgenpost	Berlin	Left of centre	182 000
Frankfurter Allgemeine	Frankfurt	Right of centre	350 000
Frankfurter Rundschau	Frankfurt	Social democratic	196 000
Die Welt	Bonn	Christian democratic	210 000
Suddeutsche Zeitung	München	Social democratic	370 000

(*b*) Radio and television: *Arbeitsgemeinschaft der öffentlich-rechtlichen Rundfunkanstalten der Bundesrepublik Deutschland* (ARD) is the co-ordinating body of the Federal German Radio and Television organizations (nine autonomous regional organizations as well as broadcast programmes for Europe and overseas).

Economic Interest Organizations:
Employers' organization: Bundesvereinigung der Deutschen Arbeitsgeberverbände.

Central trade union peak association: Deutscher Gewerkschaftsbund (DGB), first founded 1868, reorganized 1949; membership: 7,600,000 (1985).

Central Statistical Office:
Statistiches Bundesamt,
Gustav-Stresemann-Ring 11,

Postfach 5528,
D-6200 Wiesbaden 1

Further Information:

Beyme, K. von (1979), *Das politische System der Bundesrepublik Deutschland: Eine Einführung*. München: Piper.

Dalton, R. J. (1988), *Politics in West Germany*. Boston: Little, Brown.

Padgett, S., and Burkett, T. (1987), *Political Parties and Elections in Germany: The Search for a New Stability*. London: Hurst.

Smith, G. (1986), *Democracy in Western Germany: Parties and Politics in the Federal Republic*. Aldershot: Avebury.

7. Greece

State Structure: Unitary.
Form of State: Republic.
Parliament (unicameral): Parliament (Vouli): 300 seats, 4 years.
Electoral System: Proportional representation: Hagenbach quota.
Main Language: Greek (96%).
Constitutional Development: Present constitution dates from 1975 introducing a republican constitution. The military governed Greece from April 1967 to July 1974.

Heads of State:

King Georg II	Sept. 1946–Apr. 1947
King Paul I	Apr. 1947–Mar. 1964
King Kostantinos II	Mar. 1964–Dec. 1967
K. Zoitakis	Dec. 1967–Mar. 1972
G. Papadopoulos	Mar. 1972–Nov. 1973
F. Gizikis	Nov. 1973–Dec. 1974
M. Stasinopoulos	Dec. 1974–June 1975
C. Tsatsos	June 1975–May 1980
K. Karamanlis	May 1980–Mar. 1985
C. Sartzetakis	Mar. 1985–May 1989
K. Karamanlis	May 1989–

Capital City: Athínai (1981): 885,737.

Ministries

Ministry to the President
Ministry to the Prime Minister

Agriculture	Health, Welfare, and Social Security
Commerce	Industry, Energy, and Technology
Communications	Interior
Culture and Sciences	Justice
Defence	Labour
Education and Religion	Merchant Shipping
Finance	National Economy
Foreign Affairs	National Resources

Ministry to the President
Ministry to the Prime Minister

Northern Greece   Public Order
Physical Planning, Housing and   Public Work
  Environment

Governments:

	Prime Minister	Parties represented in government and distribution of ministers
Apr. 1946–Apr. 1946	Poulitsas (Indep.)	
Apr. 1946–Jan. 1947	Tsaldaris I (Prog.)	
Jan. 1947–Aug. 1947	Maximos (Indep.)	
Aug. 1947–Sept. 1947	Tsaldaris II (Prog.)	
Sept. 1947–June 1949	Sophoulis (Lib.)	
June 1949–Jan. 1950	Diomidis (Lib.)	
Jan. 1950–Mar. 1950	Theotokis (Indep.)	
Mar. 1950–Apr. 1950	Venizelos I (Lib.)	
Apr. 1950–Aug. 1950	Plastiras I (Prog.)	
Aug. 1950–Sept. 1950	Venizelos II (Lib.)	
Sept. 1950–Oct. 1951	Venizelos III (Lib.)	Lib. (8), People's (8), Soc. Dem. (6)
Oct. 1951–Oct. 1952	Plastiras II (Prog.)	Prog. (10), Lib. (6)
Oct. 1952–Nov. 1952	Kioussopoulos (Indep.)	Indep. (caretaker)
Nov. 1952–Oct. 1955	Papagos (ERE)	ERE (all)
Oct. 1955–Feb. 1956	Karamanlis I (ERE)	ERE (all)
Feb. 1956–Mar. 1958	Karamanlis II (ERE)	ERE (all)
Mar. 1958–May 1958	Georgakopoulos (Indep.)	Indep. (caretaker)
May 1958–Nov. 1961	Karamanlis III (ERE)	ERE (all)
Nov. 1961–June 1963	Karamanlis IV (ERE)	ERE (all)
June 1963–Nov. 1963	Pipinellis (ERE)	Indep. (caretaker)
Nov. 1963–Feb. 1964	Papandreou I (EDHIK)	EDHIK (all)
Feb. 1964–July 1965	Papandreou II (EDHIK)	EDHIK (all)
July 1965–Aug. 1965	Athanasiadis-Novas (EDHIK)	EDHIK (all)
Aug. 1965–Sept. 1965	Tsirimokos (EDHIK)	EDHIK (all)
Sept. 1965–Dec. 1966	Stephanopoulos (EDHIK)	EDHIK (15), ERE (1)
Dec. 1966–Apr. 1967	Paraskevopoulos (Indep.)	Indep. (all)
Apr. 1967–Dec. 1967	Kollias	Civilian

	Prime Minister	Parties represented in government and distribution of ministers
Dec. 1967–Aug. 1971	Papadopoulos I	Military
Aug. 1971–July 1972	Papadopoulos II	Military
July 1972–Oct. 1973	Papadopoulos III	Military
Oct. 1973–Nov. 1973	Markezinis	Civilian
Nov. 1973–July 1974	Androutsopoulos	Civilian
July 1974–Oct. 1974	Karamanlis V (ND)	ND (5), EDHIK (2), Indep. (12)
Oct. 1974–Nov. 1974	Karamanlis VI (ND)	Indep. (caretaker)
Nov. 1974–Nov. 1977	Karamanlis VII (ND)	ND (all)
Nov. 1977–May 1980	Karamanlis VIII (ND)	ND (all)
May 1980–Oct. 1981	Rallis (ND)	ND (all)
Oct. 1981–June 1985	Papandreou I (Pasok)	Pasok (all)
June 1985–June 1989	Papandreou II (Pasok)	Pasok (all)
June 1989–Oct. 1989	Tzannnetakis (ND)	ND, KKE
Oct. 1989–	Zolotas (Indep.)	ND, Pasok, KKE

Regions (Provinces):

	Population 1981
Greater Athens	3 027 331
Rest of Central Greece and Euboea	1 099 841
Peloponnesos	1 012 528
Ionian Islands	182 651
Epirus	324 541
Thessaly	695 654
Macedonia	2 120 481
Thrace	345 220
Aegean Islands	428 533
Crete	502 165

Media:
(*a*) Major newspapers:

	Location	Orientation	Daily circulation
Apogevmatini	Athens	Independent	130 000
Ta Nea	Athens	Liberal	155 000

	Location	Orientation	Daily circulation
Vradyni	Athens	Right-wing	72 000
Avgi	Athens	Left	55 000
Akropolis	Athens	Conservative	51 000
Rizospastis	Athens	KKE	48 000

(*b*) Radio and television: *Elliniki Radiophonia Tileorasi* (Hellenic National Radio Television) has been state controlled since 1939 and operates radio and television.

Economic Interest Organizations:
Employers' organization: Union of Greek Industrialists (EEB)
Central trade union peak association: General Confederation of Greek Workers (GSEE), founded 1918; membership: 570,000 (1984).

Central Statistical Office:
National Statistical Service of Greece
Odos Likourgou 14-16
Athens

Further Information:
Clogg, R. (1979), *A Short History of Modern Greece*. Cambridge: Cambridge U. P.
—— (1983) (ed.), *Greece in the 1980s*. London: Macmillan.
Featherstone, K., and Katsoudas, D. (1987) (eds.), *Political Change in Greece: Before and after the Colonels*. London: Croom Helm.
Mavrogordatos, G. (1983), *Stillborn Republic*. Berkeley, Calif.: University of California Press.
Mouzelis, N. (1978), *Modern Greece: Facets of Underdevelopment*. London: Macmillan.

8. Iceland

State Structure: Unitary.

Form of State: Republic.

Parliament (unicameral): Alþingi: 60 seats, 4 years. Two-chamber substitute: one-third of lower house elected by Alþingi to constitute upper house.

Electoral System: Proportional representation: d'Hondt divisor.

Main Language: Icelandic (97%).

Constitutional Development: Its present constitution dates from 1944, the year of its independence.

Heads of State:

S. Björnssen	June 1944–July 1952
A. Asgeirsson	July 1952–Aug. 1968
K. Eldjarn	Aug. 1968–Aug. 1980
V. Finnbogadottir	Aug. 1980–

Capital City: Reykjavik (1986): 91,394.

Ministries:

Office of the Prime Minister

Agriculture	Foreign Affairs
Commerce	Health and Social Security
Communications	Industry
Education	Justice and Ecclesiastical Affairs
Finance	Social Affairs
Fisheries	

Governments:

	Prime Minister	Parties represented in government and distribution of ministers
Oct. 1944–Feb. 1947	Thors II (IP)	IP (2), SDP (2), SP (2)
Feb. 1947–Dec. 1949	Stefansson (SDP)	SDP (2), IP (2),

	Prime Minister	Parties represented in government and distribution of ministers
		PP (2)
Dec. 1949–Mar. 1950	Thors III (IP)	IP (all)
Mar. 1950–Sept. 1953	Steinthorsson (PP)	PP (3), IP (4)
Sept. 1953–July 1956	Thors IV (IP)	IP (3), PP (3)
July 1956–Dec. 1958	Jonasson (PP)	PP (2), PA (2), SDP (2)
Dec. 1958–Nov. 1959	Jonsson (SDP)	SDP (all)
Nov. 1959–Nov. 1963	Thors V (IP)	IP (4), SDP (2)
Nov. 1963–July 1970	Benediktsson (IP)	IP (4), SDP (3)
July 1970–June 1971	Hafstein (IP)	IP (4), SDP (3)
June 1971–Aug. 1974	Johannesson I (PP)	PP (3), PA (2), ULL (2)
Aug. 1974–Aug. 1978	Hallgrimsson (IP)	IP (4), PP (4)
Aug. 1978–Oct. 1979	Johannesson II (PP)	PP (3), SDP (3), PA (3)
Oct. 1979–Feb. 1980	Grondal (SDP)	SDP (all)
Feb. 1980–May 1983	Thoroddsen (IP)	IP (3), PP (4), PA (3)
May 1983–July 1987	Hermannsson (PP)	IP (6), PP (4)
July 1987–Sept. 1988	Palsson (IP)	IP (4), PP (4), SDP (3)
Sept. 1988–	Hermannsson (PP)	PP (3), SDP (3), PA (3)

Regions (Sýslur):

	Population 1986
Austurland	13 131
Norðurland eystra	25 764
Norðurland vestra	10 676
Rekjavíkursvæði og Reykjanessvæði	148 883
Suðurland	20 065
Vestfirðir	10 193
Vesturland	14 940

Media:
(*a*) Major newspapers:

	Location	Orientation	Daily circulation
Morgunblaðið	Reykjavik	Independent	46 000
Dagblaðið Vísir (DV)	Reykjavik	Independent	39 000
Tíminn	Reykjavik	Progressive Party	10 000
Þjóðviljinn	Reykjavik	Socialist	6 000
Alþýðublaðið	Reykjavik	Social Democrat	4 000

(*b*) Radio and Television: *Ríkisútvarpið* (Icelandic State Broadcasting Service) ran radio and television until 1986. In 1986 its monopoly was abolished. There is now a privately owned TV station, Station Two, and several private radio stations.

Economic Interest Organizations:
Employers' organization: Vinnuveitendasamband Íslands (Employers' Federation).

Central trade union peak organization: Alþýðussamband Íslands (Icelandic Federation of Labour) (ASI), founded 1918; membership: 58,000 (1984).

Central Statistical Office:
Statistical Bureau of Iceland
Reykjavik

Further Information:
Griffiths, J. C. (1969), *Modern Iceland*. London: Pall Mall.
Iceland 1986 (1986), The Central Bank of Iceland.
Tomasson, R. F. (1980), *Iceland: The First New Society*. Minneapolis: University of Minnesota Press.

9. Ireland

State Structure: Unitary.

Form of State: Republic.

Parliament (bicameral): Oireachtas: Seanad Éireann (Upper house): 60 seats, 5 years; Dáil Éireann (Lower house): 166 seats, 5 years.

Electoral System: Proportional representation: single transferable vote.

Main Languages: English (95%); Irish (5%).

Constitutional Development: The original constitution dates from 1922 and it was renewed in 1937. From 1937 presidents have been elected in Ireland.

Heads of State:

S. T. O'Kelly	June 1945–June 1959
E. de Valera	June 1959–June 1973
E. Childers	June 1973–Nov. 1974
C. O'Dalaigh	Dec. 1974–Oct. 1976
P. Hillery	Oct. 1976–

Capital City: Dublin (1981): 915,115.

Ministries

Office of the President
Office of the Prime Minister

Agriculture	Gaeltacht
Communications	Health
Defence	Industry and Commerce
Education	Justice
Energy	Labour
Environment	Public Service
Finance	Social Welfare
Foreign Affairs	Tourism, Fisheries, and Forestry

Governments:

	Prime Minister	Parties represented in government and distribution of ministers
June 1944–Feb. 1948	de Valera I (FF)	FF (all)
Feb. 1948–May 1951	Costello I (FG)	FG (6), CP (2), CT (1),
May 1951–June 1954	de Valera II (FF)	FF (all)
June 1954–Mar. 1957	Costello II (FG)	FG (8), Lab. (4), CT (1)
Mar. 1957–June 1959	de Valera III (FF)	FF (all)
June 1959–Oct. 1961	Lemass I (FF)	FF (all)
Oct. 1961–Apr. 1965	Lemass II (FF)	FF (all)
Apr. 1965–Nov. 1966	Lemass III (FF)	FF (all)
Nov. 1966–July 1969	Lynch I (FF)	FF (all)
July 1969–Mar. 1973	Lynch II (FF)	FF (all)
Mar. 1973–July 1977	Cosgrave (FG)	FG (10), Lab. (5)
July 1977–Dec. 1979	Lynch III (FF)	FF (all)
Dec. 1979–June 1981	Haughey I (FF)	FF (all)
June 1981–Mar. 1982	Fitzgerald I (FG)	FG (11), Lab. (4)
Mar. 1982–Dec. 1982	Haughey II (FF)	FF (all)
Dec. 1982–Mar. 1987	Fitzgerald II (FG)	FG (11), Lab. (4)
Mar. 1987–July 1989	Haughey III (FF)	FF (all)
July 1989–	Haughey IV (FF)	FF, PD

Provinces:

	Population 1986
Connaught	430 726
Leinster	1 851 134
Munster	1 019 694
Ulster (part)	235 461

Media:
(a) Major newspapers:

	Location	Orientation	Daily circulation
Irish Independent	Dublin	Pro-FG	175 000
Evening Press	Dublin	Pro-FF	130 000

	Location	Orientation	Daily circulation
Evening Herald	Dublin	Pro-FG	132 000
Irish Press	Dublin	Pro-FF	87 000
Irish Times	Dublin	Independent	85 000

(*b*) Radio and television: *Raidio Telefís Éireann* (RTE) is the Irish national broadcasting organization. RTE broadcasts on two channels (Radio 1 and Radio 2) and transmits two programmes on television (RTE-1 and RTE-2). In addition *Raidio na Gaeltachta* broadcasts in Gaelic.

Economic Interest Organizations:
Employers' organization: Federated union of employers.

Central trade union peak association: Irish Congress of Trade Unions (ICTU), founded 1894; membership: 655,000 (1984).

Central Statistical Office:
Central Statistics Office
St Stephen's Green House
Earlsfort Terrace
Dublin 2

Further Information:
Chubb, B. (1982), *The Government and Politics of Ireland*. London: Longman.
Gallagher, M. (1985), *Political Parties in the Republic of Ireland*. Manchester: Manchester U.P.
Mair, P. (1987), *The Changing Irish Party System*. London: Pinter.
Peillon, M. (1982), *Contemporary Irish Society: An Introduction*. Dublin: Gill & Macmillan.

10. Italy

State Structure: Unitary.

Form of State: Republic since 1946.

Parliament (bicameral): Chamber of deputies (Camera dei deputati): 630 seats, 5 years; Senate (Senato): 323 seats, 5 years.

Electoral System: Proportional representation: imperiali quota.

Main Languages: Italian (99%).

Constitutional Development: The present constitution dates from 1948.

Heads of State:

L. Einaudi	May 1948–Apr. 1955
G. Gronchi	Apr. 1955–May 1962
A. Segni	May 1962–Dec. 1964
G. Saragat	Dec. 1964–Dec. 1971
G. Leone	Dec. 1971–June 1978
A. Fanfani (acting)	June 1978–July 1978
S. Pertini	July 1978–July 1985
F. Cossiga	July 1985–

Capital City: Roma (1985): 2,826,488.

Ministries:

Office of the President
Office of the Prime Minister

Agriculture and Forests	Interior
Budget	Justice
Cultural Heritage	Labour and Social Security
Defence	Merchant Marine
Environment	Posts and Telecommunications
Education	Public Works
Finance	State Industries
Foreign Affairs	Tourism
Foreign Trade	Transport
Health	Treasury
Industry	

Governments:

	Prime Minister	Parties represented in government and distribution of ministers
July 1946–Feb. 1947	De Gasperi II (DC)	DC, PSI, PCI, PRI
Feb. 1947–May 1947	De Gasperi III (DC)	DC, PSI, PCI
May 1947–May 1948	De Gasperi IV (DC)	DC, PRI, PSLI, PLI
May 1948–Jan. 1950	De Gasperi V (DC)	DC (11), PSLI (3), PRI (1), PLI (2)
Jan. 1950–July 1951	De Gasperi VI (DC)	DC (13), PSLI (3), PRI (3)
July 1951–Nov. 1952	De Gasperi VII (DC)	DC (14), PRI (3)
Nov. 1952–July 1953	De Gasperi VIII (DC)	DC (14), PRI (3)
July 1953–Aug. 1953	De Gasperi IX (DC)	DC (17)
Aug. 1953–Jan. 1954	Pella (DC)	DC (all)
Jan. 1954–Feb. 1954	Fanfani I (DC)	DC (all)
Feb. 1954–Mar. 1955	Scelba I (DC)	DC (14), PSDI (4), PLI (3)
Mar. 1955–July 1955	Scelba II (DC)	DC (14), PSDI (4), PLI (3)
July 1955–Mar. 1957	Segni I (DC)	DC (14), PSDI (4), PLI (3)
Mar. 1957–May 1957	Segni II (DC)	DC (15), PSDI (4) PLI (3)
May 1957–July 1958	Zoli (DC)	DC (all)
July 1958–Feb. 1959	Fanfani II (DC)	DC (17), PSDI (4)
Feb. 1959–Mar. 1960	Segni III (DC)	DC (all)
Mar. 1960–July 1960	Tambroni (DC)	DC (all)
July 1960–Feb. 1962	Fanfani III (DC)	DC (all)
Feb. 1962–June 1963	Fanfani IV (DC)	DC (19), PSDI (3) PRI (2)
June 1963–Dec. 1963	Leone I (DC)	DC (all)
Dec. 1963–July 1964	Moro I (DC)	DC (16), PSI (6), PSDI (3), PRI (1)
July 1964–Feb. 1966	Moro II (DC)	DC (16), PSI (6), PSDI (3), PRI (1)
Feb. 1966–June 1968	Moro III (DC)	DC (16), PSI (6), PSDI (3), PRI (1)

	Prime Minister	Parties represented in government and distribution of ministers
June 1968–Dec. 1968	Leone II (DC)	DC (all)
Dec. 1968–Aug. 1969	Rumor I (DC)	DC (17), PSU (9), PRI (1)
Aug. 1969–Mar. 1970	Rumor II (DC)	DC (all)
Mar. 1970–Aug. 1970	Rumor III (DC)	DC (17), PSI (6), PSDI (3), PRI (1)
Aug. 1970–Feb. 1972	Colombo (DC)	DC (16), PSI (6), PSDI (4), PRI (1)
Feb. 1972–June 1972	Andreotti I (DC)	DC (all)
June 1972–July 1973	Andreotti II (DC)	DC (17), PSDI (5), PLI (4)
July 1973–Mar. 1974	Rumor IV (DC)	DC (17), PSI (6), PSDI (4), PRI (2)
Mar. 1974–Nov. 1974	Rumor V (DC)	DC (16), PSI (6), PSDI (4)
Nov. 1974–Feb. 1976	Moro IV (DC)	DC (20), PRI (5)
Feb. 1976–July 1976	Moro V (DC)	DC (all)
July 1976–Mar. 1978	Andreotti III (DC)	DC (all)
Mar. 1978–Mar. 1979	Andreotti IV (DC)	DC (all)
Mar. 1979–Aug. 1979	Andreotti V (DC)	DC (14), PSDI (4), PRI (2)
Aug. 1979–Apr. 1980	Cossiga I (DC)	DC (17), PSDI (4), PLI (2)
Apr. 1980–Oct. 1980	Cossiga II (DC)	DC (15), PSI (7), PRI (3)
Oct. 1980–June 1981	Forlani (DC)	DC (14), PSI (7), PSDI (3), PRI (3)
June 1981–Aug. 1982	Spadolini I (PRI)	DC (15), PSI (7), PSDI (3), PRI (2), PLI (1)
Aug. 1982–Nov. 1982	Spadolini II (PRI)	DC (15), PSI (7), PSDI (3), PRI (2), PLI (1)
Nov. 1982–Apr. 1983	Fanfani V (DC)	DC (14), PSI (7), PSDI (5), PLI (2)
Apr. 1983–Aug. 1983	Fanfani VI (DC)	DC (14), PSI (5),

	Prime Minister	Parties represented in government and distribution of ministers
Aug. 1983–Apr. 1987	Craxi (PSI)	PLI (2) DC (15), PSI (6), PSDI (3), PRI (3), PLI(2)
Apr. 1987–July 1987	Fanfani VII (DC)	DC (all)
July 1987–Apr. 1988	Goria (DC)	DC (15), PSI (8), PSDI (3), PRI (3), PLI (1)
Apr. 1988–May 1989	De Mita (DC)	DC (16), PSI (10), PRI (3), PSDI (2), PLI (1)
July 1989–	Andreotti VI (DC)	DC (17), PSI (9), PRI (2) PSDI (2), PLI (2)

Regions (Regioni):

	Population 1985	Capital
Abruzzi	1 250 057	L'Aquila
Basilicata	618 647	Potenza
Calabria	2 131 412	Catanzaro
Campania	5 651 200	Napoli
Emilia-Romagna	3 939 289	Bologna
Friuli-Venezia Giulia	1 219 526	Trieste
Lazio	5 101 641	Roma
Liguria	1 771 319	Genova
Lombardia	8 881 683	Milano
Marche	1 425 734	Ancona
Molise	333 502	Campobasso
Piemonte	4 394 312	Torino
Puglia	4 005 226	Bari
Sardegna	1 638 172	Cagliari
Sicilia	5 084 311	Palermo
Toscana	3 576 508	Firenze
Trentino-Alto Adige	878 590	Trento, Bolzano
Umbria	816 939	Perugia

	Population 1985	Capital
Valle d'Aosta	113 714	Aosta
Veneto	4 370 533	Venezia

Media:
(*a*) Major newspapers:

	Location	Orientation	Daily circulation
Corriere della sera	Milano	Independent	633 000
La stampa	Torino	Liberal	556 000
L'unita	Roma-Milano	PCI	300 000
Il messaggero	Roma	Independent left	313 000
La repubblica	Roma	Independent left	504 000
Il giorno	Milano	Independent left	294 000
Il giornale	Milano	Independent	274 000
Avanti!	Roma	PSI	54 000
Il popolo	Roma	DC	43 000

(*b*) Radio and television: *Radiotelevisione Italia* (RAI) is a government corporation broadcasting on three radio and three television networks. Until 1976 RAI had a broadcasting monopoly. There are now 4 private nationwide networks and 30 local networks.

Economic Interest Organizations:
Employers' organization: Confederazione Generale dell'Industria Italiana (CONFINDUSTRIA).

Central trade union peak associations: Confederazione Generale Italiana del Lavoro (CGIL) (close to PCI, PSI), founded 1944; membership: 4,500,000 (1985); Confederazione Italiano dei Sindacati Lavoratori (CISL) (close to DC), founded 1948; membership: 2,980,000 (1985); Unione Italiana dei Lavoro (UIL) (close to PSDI, PRI, PSI), founded 1950; membership 1,344,000 (1985); Confederazione Italiana dei Sindacati Nazionale dei Lavoratore (CISNAL) (close to MSI), founded 1950; membership: 1,969,000 (1985).

Central Statistical Office:
Istituto Centrale di Statistica
Via Cesare Balbo 16,
I-00100 Roma

Further Information:
Clark, M. (1984), *Modern Italy, 1971-1982*. London: Longman.
Farneti, P. (1985), *The Italian Party System*. London: Pinter.

LaPalombara, J. (1987), *Democracy: Italian Style*. New Haven, Conn.: Yale U.P.

Mack Smith, D. (1969), *Italy: A Modern History*. Ann Arbor, Mich.: University of Michigan Press.

Sassoon, D. (1986), *Contemporary Italy: Politics, Economics and Society since 1945*. London: Longman.

11. Luxembourg

State Structure: Unitary.

Form of State: Constitutional monarchy.

Parliament (unicameral): Chambre des députés: 60 seats, 5 years.

Electoral System: Proportional representation: Hagenbach quota.

Main Languages: Luxembourgish (70%); Portuguese (8%); Italian (6%).

Constitutional Development: The present constitution dates back to 1868 but important amendments were introduced in 1919 (universal suffrage) and 1956 (nationwide elections every five years).

Heads of State:

Grand Duchess Charlotte	Jan. 1919–Nov. 1964
Grand Duke Jean	Nov. 1964–

Capital City: Luxembourg-Ville (1981): 78,900.

Ministries:

Office of the Prime Minister

Agriculture and Viticulture	Labour
Civil Service	National Economy and
Defence	the Middle Classes
Environment	National Education and Youth
Family, Housing, and Social Solidarity	Public Works
Finance	Social Security
Foreign Affairs, Foreign Trade, and Co-operation	Sport
	Tourism
Health	Transport and Energy
Interior	Treasury
Justice	

Governments:

	Prime Minister	Parties represented in government and distribution of ministers
Mar. 1947–July 1951	Dupong I (CSP)	CSP (4), DP (3)
July 1951–Dec. 1953	Dupong II (CSP)	CSP (3), SWP (3)
Dec. 1953–Mar. 1958	Bech (CSP)	CSP (3), SWP (3)
Mar. 1958–Feb. 1959	Frieden (CSP)	CSP (4), SWP (4)
Feb. 1959–June 1964	Werner I (CSP)	CSP (4), DP (3)
June 1964–Dec. 1966	Werner II (CSP)	CSP (5), SWP (5)
Dec. 1966–Jan. 1969	Werner III (CSP)	CSP (5), SWP (5)
Jan. 1969–June 1974	Werner IV (CSP)	CSP (4), DP (4)
June 1974–July 1979	Thorn (DP)	DP (4), SWP (4)
July 1979–July 1984	Werner V (CSP)	CSP (5), DP (4)
July 1984–July 1989	Santer I (CSP)	CSP (6), SWP (6)
July 1989–	Santer II (CSP)	CSP (6), SWP (6)

Districts:

	Population 1986
Luxembourg	272 250
Diekirch	54 420
Grevenmacher	40 030

Media:
(*a*) Major newspapers:

	Location	Orientation	Daily circulation
Luxemburger Wort	Luxembourg	PCS	81 000
Tageblatt	Luxembourg	POSL	24 000
La Républicain lorrain	Luxembourg	Independent	24 000

(*b*) Radio and television: *Radio-Télé Luxembourg* (RTL) broadcasts on five radio channels and two television channels.

Economic Interest Organizations:
Employers' organization: Groupe de Liaison Patronal, with: Fédération des Industriels; Fédération des Artisans; Fédération du Commerce.

Central trade union peak association: Confédération Générale du Travail du Luxembourg (CGT), founded 1927; membership: 44,000 (1984).

Central Statistical Office:
Service Central de la Statistique et des Études Économiques
Ministère de l'Économie
19-21 boulevard Royal
2910 Luxembourg

Further Information:
Als, G. (1982), *Le Luxembourg: Situation politique, économique et sociale.* Paris: La Documentation Française.
CRISP (1975) (1980) (1985), *Grand-Duché de Luxembourg: Système et comportements électoraux: Analyse et synthèse des scrutins de 1974, 1979 et 1984.* Bruxelles.

12. The Netherlands

State Structure: Unitary.

Form of State: Constitutional monarchy.

Parliament (bicameral): Staten General Eerste Kamer (First chamber): 75 seats, 4 years; Tweede Kamer (Second chamber): 150 seats, 4 years.

Electoral System: Proportional representation: d'Hondt divisor.

Main Language: Dutch (96%).

Constitutional Development: Its present constitution only dates from 1983; its first constitution was adopted in 1814-15 and was subsequently revised and amended.

Heads of State:

Queen Wilhelmina	Nov. 1890–Sept. 1948
Queen Juliana	Sept. 1948–Apr. 1980
Queen Beatrix	Apr. 1980–

Capital Cities: Amsterdam (capital) (1986): 679,140; 's Gravenhage (seat of government) (1986): 443,961.

Ministries

Office of the Prime Minister

Agriculture and Fisheries	Home Affairs
Defence	Housing, Physical Planning,
Economic Affairs	and the Environment
Education and Science	Justice
Employment and Social Security	Transport and Public Works
Finance	Welfare, Health and
Foreign Affairs	Cultural Affairs

Governments:

	Prime Minister	Parties represented in government and distribution of ministers
July 1946–Aug. 1948	Beel I (KVP)	KVP, PVDA
Aug. 1948–Mar. 1951	Drees I (PVDA)	KVP (5), PVDA (5), CHU (1), VVD (1)
Mar. 1951–Sept. 1952	Drees II (PVDA)	KVP (6), PVDA (5), CHU (2), VVD (1)
Sept. 1952–Oct. 1956	Drees III (PVDA)	KVP (5), PVDA (5), CHU (2), ARP (2)
Oct. 1956–Dec. 1958	Drees IV (PVDA)	KVP (5), PVDA (5), CHU (1), ARP (2)
Dec. 1958–May 1959	Beel II (KVP)	KVP (6), ARP (2), CHU (2)
May 1959–Jan. 1961	De Quay I (KVP)	KVP (6), ARP (2), CHU(2), VVD (2)
Jan. 1961–July 1963	De Quay II (KVP)	KVP (6), ARP (2), CHU(2), VVD (2)
July 1963–Apr. 1965	Marijnen (KVP)	KVP (6), VVD (3), CHU(2), ARP(2)
Apr. 1965–Nov. 1966	Cals (KVP)	KVP (6), PVDA (5), ARP (3)
Nov. 1966–Apr. 1967	Zijlstra (ARP)	KVP(8), ARP (5)
Apr. 1967–July 1971	De Jong (KVP)	KVP (6), VVD (3), ARP(3), CHU(2)
July 1971–May 1973	Biesheuvel (ARP)	KVP (6), ARP (3), CHU (2), VVD (3), DS-70(2)
May 1973–Dec. 1977	den Uyl (PVDA)	PVDA (7), KVP (4), ARP (2), PPR (2), D-66 (1)
Dec. 1977–Sept. 1981	van Agt I (CDA)	CDA (10), VVD (6)
Sept. 1981–Oct. 1981	van Agt II (CDA)	CDA (6), PVDA (6), D-66 (3)
Oct. 1981–Nov. 1982	van Agt III (CDA)	caretaker
Nov. 1982–July 1986	Lubbers I (CDA)	CDA (8), VVD (6)
July 1986–Nov. 1989	Lubbers II (CDA)	CDA (9), VVD (5)
Nov. 1989–	Lubbers III (CDA)	CDA (6), PVDA (8)

Provinces:

	Population 1987	Capital
Groningen	558 378	Groningen
Friesland	599 061	Leeuwarden
Drenthe	434 038	Assen
Overijssel	1 003 915	Zwolle
Flevoland (earlier Ijsselmeerpolders)	185 365	Lelystad
Gelderland	1 771 972	Arnhem
Utrecht	953 957	Utrecht
Noord-Holland	2 334 209	Haarlem
Zuid-Holland	3 186 249	's Gravenhage
Zeeland	335 434	Middelburg
Noord-Brabant	2 139 626	's Hertogenbosch
Limburg	1 091 553	Maastricht

Media:
(a) Major newspapers:

	Location	Orientation	Daily circulation
Het Parool	Amsterdam	Independent	134 000
De Telegraaf	Amsterdam	Independent	688 000
De Volkskrant	Amsterdam	Independent	285 000
Algemeen Dagblad	Rotterdam	Independent	382 000
Weekly	Rotterdam	Socialist	197 000
Trouw	Amsterdam	Protestant	124 000
NRC/Handelsblad	Rotterdam	Independent	187 000

(*b*) Radio and television: *Nederlandse Omroep Stichting* (NOS) is the co-ordinating body for eight associate broadcasting companies transmitting programmes on radio and television. Among these the following may be named: *Algemene Omroepvereniging* (AVRO) (Independent); *Omroepvereniging Vara* (Socialist); *Katholieke Radio Omroep* (KRO) (Catholic); *Nederlands Christelijke Radio*; *Televisie Radio Omroep Stichting* (TROS) (Independent); *Vereniging* (NCRV) (Protestant); *Veronica Omroep Organisatie* (VOO) (Independent); *VPRO* (Independent).

Economic Interest Organizations:
Employers' organizations: Verbond van Nederlandse Ondernemingen (VNO); Nederlandse Christelijke Werkgeversorganisatie (NCW).
    Central trade union peak associations: Federatie Nederlandse Vakbe-

weging (FNV), formed in 1976 as a confederation of the socialist Nederlands Verbond van Vakverenigingen (NVV) and the Catholic Nederlands Katholiek Vakverboend (NKV); membership: 906,000 (1988); Christelijk Nationaal Vakverbond in Nederland (CNV), formed 1909; membership: 292,000 (1988); Volkcentrale van Middelbaar en Hoger Personeel (MHP); membership: 117,000 (1988).

Central Statistical Office:
Netherlands Central Bureau of Statistics
Prinses Beatrixlaan 428
POB 959
NL-2270 AZ Voorburg

Further Information:
Daalder, H., and Schuyt, C. (1986) (eds.), *Compendium voor Politiek en Samenleving*. Alphen aan den Rijn.
Daalder, H., and Irwin, D. (1989) (eds.), *Politics in the Netherlands: How much Change?* West European Politics, 12, London.

13. Norway

State Structure: Unitary.

Form of State: Constitutional monarchy.

Parliament (unicameral): Storting (divided into Lagting and Odelsting when it comes to legislation): 165 seats, 4 years.

Electoral System: Proportional representation: Sainte Lague, divisor 1, 4 + 8 seats to be distributed on nationwide result.

Main Language: Norwegian (98%).

Constitutional Development: The present constitution dates from 1814 but has been amended.

Heads of State:

King Haakon VII	Nov. 1905–Sept. 1957
King Olav V	Sept. 1957–

Capital City: Oslo   (1986): 449,395.

Ministries

Office of the Prime Minister

Agriculture	Environment
Business and Industry	Finance
Church and Education	Fisheries
Communications	Foreign Affairs
Consumer Affairs and Government Administration	Health and Social Affairs
	Justice
Cultural and Scientific Affairs	Local Government and Labour
Defence	Petroleum and Energy
Development Co-operation	

Governments:

	Prime Minister	Parties represented in government and distribution of ministers
Nov. 1945–Nov. 1951	Gerhardsen II (DNA)	DNA (all)
Nov. 1951–Jan. 1955	Torp (DNA)	DNA (all)

	Prime Minister	Parties represented in government and distribution of ministers
Jan. 1955–Aug. 1963	Gerhardsen III (DNA)	DNA (all)
Aug. 1963–Sept. 1963	Lyng (H)	H(5), V (3), SP (4), KRF (3)
Sep. 1963–Oct. 1965	Gerhardsen IV (DNA)	DNA (all)
Oct. 1965–Mar. 1971	Borten (SP)	SP(3), V (3), H (6), KRF (3)
Mar. 1971–Oct. 1972	Bratteli I (DNA)	DNA (all)
Oct. 1972–Oct. 1973	Korvald (KRF)	KRF (4), SP (6), V (5)
Oct. 1973–Jan. 1976	Bratteli II (DNA)	DNA (all)
Jan. 1976–Feb. 1981	Nordli (DNA)	DNA (all)
Feb. 1981–Oct. 1981	Brundtland I (DNA)	DNA (all)
Oct. 1981–June 1983	Willoch I (H)	H (all)
June. 1983–Mar. 1986	Willoch II (H)	H (11), SP (4), KRF (3)
Mar. 1986–Oct. 1989	Brundtland II (DNA)	DNA (all)
Oct. 1989–	Syse (H)	H (9), SP (5), KRF (5)

Counties (Fylker):

	Population 1986	Capital
Østfold	234 989	Moss
Akershus	393 239	—
Oslo	449 395	Oslo
Hedmark	186 366	Hamar
Oppland	181 796	Lillehammer
Buskerud	219 993	Drammen
Vestfold	191 570	Tønsberg
Telemark	162 529	Skien
Aust-Agder	94 669	Arendal
Vest-Agder	140 212	Kristianssand
Rogaland	323 310	Stavanger
Hordaland	399 662	Bergen
Sogn og Fjordane	105 966	Leikanger
Møre og Romsdal	237 230	Molde

	Population 1986	Capital
Sør-Trøndelag	246 800	Trondheim
Nord-Trøndelag	126 711	Stenkjær
Nordland	242 240	Bodø
Troms	146 702	Tromsø
Finnmark	75 668	Vardø

Media:
(a) Major newspapers:

	Location	Orientation	Daily circulation
Aftenposten	Oslo	Independent conservative	239 000
Arbeiderbladet	Oslo	DNA	57 000
Dagbladet	Oslo	Independent liberal	175 000
Nationen	Oslo	SP	20 000
Verdens Gang	Oslo	Independent	290 000
Vårt Land	Oslo	KRF	26 000

(b) Radio and television: *Norsk Rikskringkastning* (NRK) is the broadcasting board of the state.

Economic Interest Organizations:
Employers' organization: As of 1988, NHO, Næringslivets Hovedorganisasjon (Confederation of Norwegian Business and Industry).

Central trade union peak associations: Landsorganisationen i Norge (LO), founded 1899; membership: 785,573 (1986); Confederation of Vocational Unions: Yrkesorganastioners Centralsörbund; membership: 131,739 (1986); Federation of Norwegian Professional Associations: Akaadaemikernas Sallasforbund; membership: 144,108 (1986). Other unions: 170,182 (1986).

Central Statistical Office:
Statistisk Sentralbyrå
POB 8131
N-0033 Oslo 1

Further Information:
Brochmann, B. S., and Josefsen, D. (1984), *Fiskerinæringen*. Oslo: Tiden.
Derry, T. K. (1973), *A History of Modern Norway 1814-1972*. Oxford: Clarendon Press.

Eliassen, K. A., and Pederssen, M. N. (1986), *Nordiske politiske facta*. Oslo: Tiden.
Heidar, K. (1982), *Norske politiske fakta 1884-1982*. Oslo: Universitetsforlaget.
Kuhnle, S. (1983), *Velferdsstatens utvikling*. Oslo: Universitetsforlaget.
—— (1983), *Velferdsstaten*. Oslo: Tiden.
Lagreid, P., and Roness, P. (1983), *Sentraladministrasjonen*. Oslo: Tiden.
Olsen, J. P. (1983), *Organized Democracy: Political Institutions in a Welfare State: The Case of Norway*. Oslo: Universitetsforlaget.
Ostbye, H. (1984), *Massmedia*. Oslo: Tiden.
Rokkan, S. (1970), *Citizens, Elections, Parties*. Oslo: Universitetsforlaget.
Svåsand, L. (1985), *Politiske partier*. Oslo: Tiden.

14. Portugal

State Structure: Unitary.

Form of State: Republic.

Parliament (unicameral): Assembleia da República: 250 seats, 4 years.

Electoral System: Proportional representation: d'Hondt divisor.

Main Language: Portuguese (99%).

Constitutional Development: After the overthrow of the authoritarian regime the present constitution came into force on 25 April 1976. It was revised in 1982 and in 1989.

Heads of State:

A. de Spinola	May 1974–Sept. 1974
F. da Costa Gomes	Sept. 1974–June 1976
A. R. Eanes	July 1976–Mar. 1986
M. Soares	Mar. 1986 –

Capital City: Lisboa (1981): 807,937.

Ministries

Office of the President
Office of the Prime Minister

Agriculture
Education, Scientific Research, Universities and Culture
European Integration
Finance
Foreign Affairs
Health
Internal Administration
Justice
Labour and Social Security
National Defence
Parliamentary Affairs
Plan and Territory Administration
Public Works, Transport, and Communications
Quality of Life
State
Trade and Industry

Governments:

	Prime Minister	Parties represented in government and distribution of ministers
May 1974–July 1974	Palma Carlos (Indep.)	Pro-communist
July 1974–Sept. 1974	Goncalves I (Indep.)	Pro-communist
Sept. 1974–Mar. 1975	Goncalves II (Indep.)	Pro-communist
Mar. 1975–Aug. 1975	Goncalves III (Indep.)	PS (2), PCP (2), PPD (2), MDP-CDE (1), Indep. (14)
Aug. 1975–Sept. 1975	Goncalves IV (Indep.)	Pro-communist (all)
Sept. 1975–July 1976	de Azevedo (Indep.)	PS (4), PCP (1), PPD (2), Indep. (8)
July 1976–Jan. 1978	Soares I (PS)	PS (14), Indep. (6)
Jan. 1978–Aug. 1978	Soares II (PS)	PS (11), CDS (3)
Aug. 1978–Nov. 1978	Nobre da Costa (Indep.)	Indep. (all)
Nov. 1978–July 1979	Mota Pinto (Indep.)	Indep. (all)
July 1979–Jan. 1980	Pintasilgo (Indep.)	Indep. (all)
Jan. 1980–Jan. 1980	Sa Carneiro (PSD)	PSD (9), CDS (5)
Jan. 1980–Sept. 1981	Pinto Balsemao I (PSD)	PSD (10), CDS (5), PPM (1)
Sept. 1981–Jan. 1983	Pinto Balsemao II (PSD)	PSD (8), CDS (5), PPM(1)
Jan. 1983–June 1983	Pinto Balsemao III (PSD)	Caretaker
June 1983–June 1985	Soares III (PS)	PS (9), PSD (7)
June 1985–Nov. 1985	Soares IV (PS)	PS (all)
Nov. 1985–Aug. 1987	Cavaco Silva I (PSD)	PSD (all)
Aug. 1987–	Cavaco Silva II (PSD)	PSD (all)

Districts:

	Population 1981	Capital
Aveiro	622 988	Aveiro
Beja	188 420	Beja
Braga	708 924	Braga
Bragança	184 252	Bragança
Castelo Branco	234 230	Castelo Branco
Coimbra	436 324	Coimbra

	Population 1981	Capital
Évora	180 277	Évora
Faro	323 534	Faro
Guarda	205 631	Guarda
Leiria	420 229	Leiria
Lisboa	2 069 467	Lisboa
Portalegre	142 905	Portalegre
Pôrto	1 562 287	Pôrto
Santarém	454 123	Santarém
Setúbal	658 326	Setúbal
Viana do Castelo	256 814	Viana do Castelo
Vila Real	264 381	Vila Real
Viseu	423 648	Viseu
Autonomous Regions		
Açores	243 410	Ponta Delgada
Madeira	252 844	Funchal

Media:
(a) Major newspapers:

	Location	Orientation	Daily circulation
A capital	Lisboa	Independent	45 000
Correio da manhã	Lisboa	Independent	78 000
Diário popular	Lisboa	Leftist	62 000
Diário de Notícias	Lisboa	Independent	59 000
Diário de Lisboa	Lisboa	Leftist	42 000
Jornal de notícias	Lisboa	Leftist	79 000
O comércio do porto	Lisboa	Moderate	54 000

(b) Radio and television: *Radiodifusão Portuguesa* (RDP) is the state-controlled broadcasting company; in 1983 the establishment of private radio stations was permitted. *Radiotelevisão Portuguesa* (RTP) is the state-controlled television company; in 1987 private television channels were permitted.

Economic Interest Organizations:
Employers' organizations: Confederação da Indústria Portuguesa (CIP), Confederação do Comércio Portuguesa (CCP), and Confederação dos Agricultores de Portugal (CAP).
Central trade union peak associations: Confederação Geral dos

Trabalhadores Portugueses-Intersindical Nacional (CGTP-INTERSIN-DICAI) (close to PCP), founded secretly 1970, present name from 1978; claims 87% of organized labour (1984); União Geral dos Trabalhadores de Portugal (UGT) (close to PS and PSD), founded 1978; membership: 1,190,000 (1984).

Central Statistical Office:
Instituto Nacional de Estatística
Av. António José de Almeida
P-1078 Lisboa Codex

Further Information:
Bruneau, T., and Macleod, A. (1986), *Politics in Contemporary Portugal*. Boulder, Colo.: Lynne Rienner.
Gallagher, T. (1983), *Portugal: A Twentieth Century Interpretation*. Manchester: Manchester U. P.
Graham, L. and Makler, H. (1979), *Contemporary Portugal: The Revolution and its Antecedents*. Austin, Tex.: University of Texas Press.
—— and Wheeler, D. (1984), *In Search of Portugal: The Revolution and its Antecedents*. Madison, Wis.: University of Wisconsin Press.
Marques, A. (1975), *História de Portugal*, 2 vols. Lisboa: Palas Editores.
Maxwell, K. (1986), *Portugal in the 1980s*. Westport: Greenwood Press.
Baptista, C. (Co-ord.) (1989), *Portugal: O sistema político e constitucional 1974/87*. Lisboa: IPS.
Robinson, R. A. H. (1979), *Contemporary Portugal: A History*. London: Allen & Unwin.

15. Spain

State Structure: Unitary.

Form of State: Constitutional monarchy.

Parliament (bicameral): Cortes Senado: 257 seats, 4 years; Congreso de diputados: 350 seats, 4 years.

Electoral System: Proportional representation: d'Hondt divisor.

Main Languages: Castilian Spanish (73%); Catalan (16%); Galician (8%); Basque(2%).

Constitutional Development: The present constitution dates from 1978 and represents a definitive break with the Franco regime. There is a process of developing regional self-government.

Heads of State:

King Juan Carlos   Nov. 1975 –

Capital City: Madrid (1981): 3,188,297.

Ministries

Prime Minister's Chancellery

Agriculture, Fisheries, and Food	Health and Consumer Affairs
Culture	Industry and Energy
Defence	Interior
Economy, Finance, and Trade	Justice
Education and Science	Labour and Social Security
Foreign Affairs	Public Administration
	Public Works and Town Planning
	Transport, Tourism, and Communications

Governments:

	Prime Minister	Parties represented in government and distribution of ministers.
July 1976–July 1977	Suarez I (UCD)	UCD (all)
July 1977–Apr. 1979	Suarez II (UCD)	UCD (all)

	Prime Minister	Parties represented in government and distribution of ministers.
Apr. 1979–Feb. 1981	Suarez III (UCD)	UCD (all)
Feb. 1981–Dec. 1982	Calvo Sotelo (UCD)	UCD (all)
Dec. 1982–July 1986	Gonzales I (PSOE)	PSOE (15), PSC (2)
July 1986–Nov. 1989	Gonzales II (PSOE)	PSOE (all)
Nov. 1989–	Gonzales III (PSOE)	PSOE (all)

Autonomous Communities:

	Population (1986)	Capital
Andalucía	6 735 600	Sevilla
Aragón	1 215 600	Zaragoza
Asturias	1 140 100	Oviedo
Baleares	675 400	Palma de Mallorca
Canarias	1 442 500	Santa Cruz de Tenerife
Cantabria	527 400	Santander
Castilla-La Mancha	1 670 100	Toledo
Extremadura	1 084 400	Mérida
Galicia	2 870 900	Santiago de Compostela
La Rioja	263 100	Logroño
Madrid	4 907 100	Madrid
Murcia	1 007 500	Murcia
Navarra	522 500	Pamplona
País Vasco	2 176 800	Vitoria
Valencia	3 790 200	Valencia

Media:
(*a*) Major newspapers:

	Location	Orientation	Daily circulation
El país		Independent left	348 000
ABC		Independent monarchist	240 000
Ya		Catholic, right	96 000

(*b*) Radio and television: *Radiotelevisión Española* (RTVE) controls and co-ordinates Spanish radio and television. There are in addition to the state-controlled radio many commercial and independent radio stations. Legislation aiming to end the monopoly of TVE was introduced in 1987.

Economic Interest Organizations:
Employers' organization: Confederación Española de Organizaciones Empresariales (CEOE).

Central trade union peak associations: Confederación Sindical de Comisiones Obreras (CSCO) (independent, but close to the communists), founded in the 1960s; membership: 1,605,000 (1984); Confederación Nacional de Trabajo (CNT), (anarchist), founded 1910; membership: 150,000 (1984); Unión General de Trabajadores (UGT) (close to PSOE), founded 1888, reorganized 1977; membership: 700,000 (1984).

Central Statistical Office:
Instituto Nacional de Estadística
Paseo Castellana 183
E-28071 Madrid

Further Information:
Carr, R. (1980), *Modern Spain 1875–1980*. Oxford: Oxford U.P.
Coverdale, J. F. (1979), *The Political Transformation of Spain after Franco*. New York: Praeger.
Donaghy, P. J. and Newton, M. Y. (1987), *Spain: A Guide to Political and Economic Institutions*. Cambridge: Cambridge U.P.
Gilmour, D. (1985), *The Transformation of Spain: From Franco to the Constitutional Monarchy*. London: Quartet.
Gunther, R., *et al.* (1986), *Spain After Franco: The Making of a Competitive Party System*. Berkeley, Calif.: University of California Press.
Preston, P. (1986), *The Triumph of Democracy in Spain*. London: Methuen.
Share, D. (1987), *The Making of Spanish Democracy*. New York: Praeger.

16. Sweden

State Structure: Unitary.

Form of State: Monarchy.

Parliament (unicameral since 1970): Riksdag: 349 seats; 3 years.

Electoral System: Proportional representation: Sainte Lague divisor; use of a national constituency for allocating 39 equalization seats.

Main Language: Swedish (97%); Finnish (3%).

Constitutional Development: The present constitution dates from 1974 when the one from 1809 was replaced.

Heads Of State:

King Gustav V	Dec. 1907–Oct. 1950
King Gustav VI	Dec. 1950–Sept. 1973
King Karl XVI Gustav	Sept. 1973 –

Capital City: Stockholm (1985): 659,030.

Ministries

Office of the Prime Minister

Agriculture	Housing and Physical Planning
Defence	Industry
Education and Cultural Affairs	Justice
Environment and Energy	Labour
Finance	Public Administration
Foreign Affairs	Transport and Communications
Health and Social Affairs	

Governments:

	Prime Minister	Parties represented in government and distribution of ministers
Jan. 1945–Oct. 1946	Hansson (SAP)	SAP (all)
Oct. 1946–Sep. 1951	Erlander I (SAP)	SAP (all)
Sept. 1951–Oct. 1957	Erlander II (SAP)	SAP (11), BF (4)
Oct. 1957–Oct. 1969	Erlander III (SAP)	SAP (all)

Sweden

	Prime Minister	Parties represented in government and distribution of ministers
Oct. 1969–Oct. 1976	Palme I (SAP)	SAP (all)
Oct. 1976–Oct. 1978	Fälldin I (CP)	CP (8), M (6), FP (5)
Oct. 1978–Oct. 1979	Ullsten (FP)	FP (all)
Oct. 1979–May 1981	Fälldin II (CP)	CP (6), M (8), FP (5)
May 1981–Oct. 1982	Fälldin III (CP)	CP (10), FP (7)
Oct. 1982–Mar. 1986	Palme II (SAP)	SAP (all)
Mar. 1986–	Carlsson (SAP)	SAP (all)

Counties:

	Population 1985	Capital
Stockholm	1 578 299	Stockholm
Uppsala	251 852	Uppsala
Södermanland	249 701	Nyköping
Östergötland	393 585	Linköping
Jönköping	300 753	Jönköping
Kronoberg	173 972	Växjö
Kalmar	238 176	Kalmar
Gotland	56 144	Visby
Blekinge	150 959	Karlskrona
Kristianstad	280 354	Kristianstad
Malmöhus	750 140	Malmö
Halland	240 063	Halmstad
Göteborgs and Bohuslän	715 728	Göteborg
Älvsborg	426 698	Vänersborg
Skaraborg	270 468	Mariestad
Värmland	279 183	Karlstad
Örebro	270 211	Örebro
Västmanland	254 761	Västerås
Kopparberg	283 880	Falun
Gävleborg	289 153	Gävle
Västernorrland	262 314	Härnösand
Jämtland	134 190	Östersund
Västerbotten	245 255	Umeå

	Population 1985	Capital
Norrbotten	262 300	Luleå

Media:
(a) Major newspapers:

	Location	Orientation	Daily circulation
Expressen	Stockholm	Liberal (FP)	577 000
Dagens Nyheter	Stockholm	Independent liberal	396 000
Aftonbladet	Stockholm	SAP	377 000
Svenska Dagbladet	Stockholm	Independent moderate	224 000

(b) Radio and television: *Sveriges Radio* (Swedish Broadcasting Corporation) operates non-commercial radio and television under licence from the state. There are three radio and two television networks.

Economic Interest Organizations:
Employers' organization: Svenska Arbetsgivareföreningen (SAF).

Central trade union peak associations: Landsorganisationen i Sverige (LO) (Swedish Trade Union Confederation), founded 1898; membership: 2,263,000 (1985); Tjänstemännens Centralorganisation (TCO) (Central Organization of Salaried Employees), present name 1944; membership: 1,211,000 (1985); Centralorganisation SACO/SR (Confederation of Professional Associations), present form 1975; membership: 216,000 (1985).

Central Statistical Office:
Statistiska Centralbyrån
S-115 81 Stockholm

Further Information:
Back, P. E., and Berglund, S. (1978), *Det svenska partiväsendet*. Stockholm: Almqvist & Wiksell.
Hancock, M. D. (1972), *The Politics of Post-industrial Change*. Hinsdale: Dryden Press.
Heclo, H., and Madsen, H. (1987), *Policy and Politics in Sweden*. Philadelphia: Temple U.P.
Holmberg, S., and Gilljam, M. (1987), *Väljare och val i Sverige*. Stockholm: Bonniers.
—— and Esaiasson, P. (1988), *De folkvalda: En bok om riksdagsledamöter-*

na och den representativa demokratin i Sverige. Stockholm: Bonniers.
Koblik, S. (1975) (ed.), *Sweden's Development from Poverty to Affluence 1750–1970*. Minneapolis: University of Minnesota Press.
Lewin, L. (1989), *Ideology and Strategy: A Century of Swedish Politics*. Cambridge: Cambridge U. P.

17. Switzerland

State Structure: Federal.

Form of State: Republic.

Parliament (bicameral): Federal assembly (Bundesversammlung): Ständerat: 46 seats, 4 years; Nationalrat: 200 seats, 4 years.

Electoral System: Proportional representation: Hagenbach quota.

Main Languages: German (65%); French (18%); Italian (10%); Romansh (1%).

Constitutional Development: The present constitution dates from 1874. Attempts at drafting a new constitution in the 1980s were vetoed.

Heads of State:

The Presidents are elected on a yearly basis:

1945 E. von Steiger	1960 M. Petitpierre	1975 P. Graber
1946 K. Kobelt	1961 F. T. Wahlen	1976 R. Gnägi
1947 M. Petitpierre	1962 P. Chaudet	1977 K. Furgler
1948 E. Celio	1963 W. Spühler	1978 W. Ritschard
1949 E. Nobs	1964 L. von Moos	1979 H. Hürlimann
1950 M. Petitpierre	1965 H. Tschudi	1980 G. A. Chevallaz
1951 E. von Steiger	1966 H. Schattner	1981 K. Furgler
1952 K. Kobelt	1967 R. Bonvin	1982 F. Honegger
1953 P. Etter	1968 W. Spühler	1983 P. Aubert
1954 R. Rubattel	1969 L. von Moos	1984 L. Schlumpf
1955 M. Petitpierre	1970 H. Tschudi	1985 K. Furgler
1956 M. Feldmann	1971 R. Gnägi	1986 A. Elgli
1957 H. Streuli	1972 N. Celio	1987 P. Aubert
1958 T. Holenstein	1973 R. Bonvin	1988 O. Stich
1959 P. Chaudet	1974 E. Brugger	

Capital City: Bern   (1985): 301,100.

Ministries

Finance
Foreign Affairs
Home Affairs
Justice and Police
Federal Chancellery
Public Economy
Transport, Communications, and Energy
Military Department

Governments:

	Prime Minister	Parties represented in government and distribution of ministers
Oct. 1947–Oct. 1951	Petitpierre I (FDP)	FDP (3), CVP (2), SVP (1), SPS (1)
Oct. 1951–Oct. 1953	Kobelt (FDP)	FDP (3), CVP (2), SVP (1), SPS (1)
Oct. 1953–Oct. 1955	Rubattel (FDP)	FDP (4), CVP (2), SVP (1)
Oct. 1955–Oct. 1959	Feldmann (SVP)	FDP (3), CVP (3), SVP (1)
Oct. 1959–Oct. 1963	Petitpierre II (FDP)	FDP (2), CVP (2), SPS (2), SVP (1)
Oct. 1963–Oct. 1967	von Moos (CVP)	FDP (2), CVP (2), SPS (2), SVP (1)
Oct. 1967–Oct. 1971	Spühler (SPS)	FDP (2), CVP (2), SPS (2), SVP (1)
Oct. 1971–Oct. 1975	Celio (FDP)	FDP (2), CVP (2), SPS (2), SVP (1)
Oct. 1975–Oct. 1979	Gnägi (SVP)	FDP (2), CVP (2), SPS (2), SVP (1)
Oct. 1979–Oct. 1983	Chevallaz (FDP)	FDP (2), CVP (2), SPS (2), SVP (1)
Oct. 1983–Oct. 1987	Aubert (SPS)	FDP (2), CVP (2), SPS (2), SVP (1)
Oct. 1987–	Stich (SPS)	FDP (2), CVP (2), SPS (2), SVP (1)

Cantons:

	Population 1980	Capital
Zürich	1 122 839	Zürich
Bern	912 022	Bern
Luzern	296 159	Luzern
Uri	33 883	Altdorf
Schwyz	97 354	Schwyz
Obwalden	25 865	Sarnen
Nidwalden	28 617	Stans

	Population 1980	Capital
Glarus	36 718	Glarus
Zug	75 930	Zug
Fribourg	185 246	Fribourg
Solothurn	218 102	Solothurn
Basel–Stadt	203 915	Basel
Basel–Land	219 822	Liestal
Schaffhausen	69 413	Schaffhausen
Appenzell Ausserrhoden	47 611	Herisau
Appenzell Innerrhoden	12 844	Appenzell
St Gallen	391 955	St Gallen
Graubünden	164 641	Chur
Aargau	453 442	Aarau
Thurgau	183 795	Frauenfeld
Ticino	265 899	Bellinzona
Vaud	528 747	Lausanne
Valais	218 707	Sion
Neuchâtel	158 368	Neuchâtel
Genève	349 040	Genève
Jura	64 986	Delémont

Media:
(*a*) Major newspapers:

	Location	Orientation	Daily circulation
Blick (Zürich)	Zürich	Independent	380 000
Tages Anzeiger Zürich	Zürich	Independent	257 000
Neue Zürcher Zeitung	Zürich	Independent liberal	145 000
Journal de Genève	Genève	Independent liberal	20 000

(*b*) Radio and television: *Schweizerische Radio- und Fernsehgesellschaft* (SRG) is an autonomous corporation under federal supervision responsible for the programme services. There are programmes on radio and television for the different language groups: Italian, French, German, and Romansh. Limited direct advertising is allowed.

Economic Interest Organizations:
Employers' organization: Zentralverband schweizerischer Arbeitgeber-Organisationen.
   Central trade union peak associations: Schweizerischer Gewerk-

schaftsbund (SGB), founded 1880; membership: 459,000 (1984); Christlichnationaler Gewerkschaftsbund der Schweiz (CNG), founded 1907; membership: 106,937 (1984).

Central Statistical Office:
Federal Office of Statistics
Hallwylstr. 15
CH-3003 Bern

Further Information:

Blancpain, R. (1978) (ed.), *Almanach der Schweiz: Daten und Kommentare zu Bevölkerung, Gesellschaft und Politik.* Bern: Peter Lang.

Gruner, E. (1971) (ed.), *Schweiz seit 1945.* Bern: Francke.

Katzenstein, P. (1980), *Capitalism in One Country? Switzerland in the International Economy.* Ithaca, NY: Cornell U. P.

Kerr, H. H. (1974), *Switzerland: Social Cleavages and Partisan Conflict.* London: Sage.

McRae, K. D. (1983), *Conflict and Compromise in Multilingual Societies: Switzerland.* Waterloo, Ont.: Wilfrid Laurier U. P.

Schmidt, M. G. (1985), *Der Schweizerische Weg zur Vollbeschäftigung.* Frankfurt am Main: Campus.

Sidjanski, D., *et al.* (1975), *Les Suisses et la politique.* Bern: Herbert Lang.

Steinberg, J. (1980), *Why Switzerland?* Cambridge: Cambridge U. P.

18. Turkey

State Structure: Unitary.

Form of State: Republic since 1922.

Parliament (unicameral): Grand national assembly (Turkiye Büjük Millet Meclisi): 400 seats, 5 years.

Electoral System: Plurality formula.

Main Languages: Turkish (86%); Kurdish (11%); Arabic 1(%).

Constitutional Development: The present constitution dates from 1982 replacing the one from 1961. The military took over in October 1981 in the form of a National Security Council. More democratic measures were being introduced in the late 1980s.

Heads of State:

Ismet Inönü	Nov. 1938–May 1950
Celal Bayar	May 1950–May 1960
Cemal Gürsel	May 1960–Mar. 1966
Cevdet Sunay	Mar. 1966–Apr. 1973
Fahri Korüturk	Apr. 1973–Apr. 1980
Ihsan Sabri Caglayangil	Apr. 1980–Sept. 1980
Kenan Evren	Sept. 1980–Nov. 1989
Turgut Özal	Nov. 1989–

Capital City: Ankara (1980): 1,877,755.

Ministries

Office of the President
Office of the Prime Minister
Office of the Deputy Prime Minister

Agriculture, Forestry, and Rural Affairs	Justice
Culture and Tourism	Labour and Social Security
Energy and Natural Resources	National Defence
Finance and Customs	National Education, Youth, and Sports
Foreign Affairs	
Health and Social Welfare	Public Works and Housing
Industry and Commerce	Transport and Communications
Interior	

Governments:

	Prime Minister	Parties represented in government
Aug. 1946–Sept. 1947	Peker (People's)	People's
Sept. 1947–June 1948	Saka I (People's)	People's
June 1948–Jan. 1949	Saka II (People's)	People's
Jan. 1949–May 1950	Günaltay (People's)	People's
May 1950–Mar. 1951	Menderes I (Dem.)	Dem.
Mar. 1951–May 1954	Menderes II (Dem.)	Dem.
May 1954–Nov. 1955	Menderes III (Dem.)	Dem.
Nov. 1955–Nov. 1957	Menderes IV (Dem.)	Dem.
Nov. 1957–May 1960	Menderes V (Dem.)	Dem.
May 1960–Jan. 1961	Gürsel I	Military
Jan. 1961–Oct. 1961	Gürsel II	Military
Oct. 1961–June 1962	Ismet VI (RRP)	RRP, JP
June 1962–Dec. 1963	Ismet VII (RRP)	RRP, NTP, CNP
Dec. 1963–Feb. 1965	Ismet VIII (RRP)	RRP
Feb. 1965–Oct. 1965	Ürgüplü (JP)	JP, RRP, NTP, CNP
Oct. 1965–Nov. 1969	Demirel I (JP)	JP
Nov. 1969–Mar. 1970	Demirel II (JP)	JP
Mar. 1970–Mar. 1971	Demirel III (JP)	JP
Mar. 1971–Dec. 1971	Erim I (Indep.)	JP, RRP
Dec. 1971–May 1972	Erim II (Indep.)	JP, RRP
May 1972–Apr. 1973	Melen (NRP)	NRP, RRP, JP
Apr. 1973–Jan. 1974	Talu (Indep.)	RRP, JP
Jan. 1974–Nov. 1974	Ecevit I (RPP)	RPP, NSP
Nov. 1974–Mar. 1975	Irmak (Indep.)	RPP
Mar. 1975–June 1977	Demirel IV (JP)	JP, NSP, NAP
June 1977–July 1977	Ecevit II (RPP)	RPP
July 1977–Jan. 1978	Demirel V (JP)	JP, NSP, NAP
Jan. 1978–Nov. 1979	Ecevit III (RPP)	RPP, RRP, DP
Nov. 1979–Sept. 1980	Demirel VI (JP)	JP, HAP, NSP
Sept. 1980–Nov. 1983	Ülüsil	Military
Nov. 1983–Nov. 1989	Özal (ANAP)	ANAP
Nov. 1989–	Akbulut (ANAP)	ANAP

Regions:

	Population 1985
Mediterranean Coast	4 653 426
West Anatolia	3 538 253
East Anatolia	6 290 086
South-east Anatolia	2 413 593
Central Anatolia	12 193 155
Black Sea Coast	6 652 172
Marmara and Aegean Coasts	9 834 576
Thrace	5 089 197

Media:
(a) Major newspapers:

	Location	Daily circulation
Cumhuriyet	Istanbul	110 000
Milliyet	Istanbul	210 000
Hürriyet	Istanbul	600 000

(b) Radio and television: *Türkiye Radyo Televizyon Kurumu* (TRT) controls Turkish radio and television services. TRT is an autonomous public corporation funded by licence fees.

Economic Interest Organizations:
Employers' organization: Türkiye İşveren Sendikaları Konfederasyonu (TISK) (Turkish Confederation of Employers' Associations).

Central trade union peak association: DISK (Confederation of Progressive Trade Unions of Turkey), founded 1967, suspended in September 1980; membership: 600,000 (1980); Türk-Is (Confederation of Turkish Trade Unions), founded 1952; membership: 1,947,000 (1985).

Further Information:
Caglar, Keyder (1987), *State and Class in Turkey*. London: Verso.
Dodd, C. H. (1969), *The Politics and Government of Turkey*. Manchester: Manchester U. P.
Heper, M., and Evin, A. (1988) (eds.), *State, Democracy and the Military: Turkey in the 1980s*. Berlin: de Gruyter.

19: United Kingdom

State Structure: Unitary.

Form of State: Constitutional monarchy.

Parliament (bicameral): House of Lords: 1,180 seats, continuous; House of Commons: 650 seats, 5 years (max).

Electoral System: Plurality formula.

Main Languages: English (93%); Welsh (1%).

Constitutional Development: There is no written constitution, but documents forming the core of a would-be constitution could be: Magna Carta of 1215, Bill of Rights of 1689, and the Reform Bill of 1832.

Heads of State:

King George VI	Dec. 1936–Feb. 1952
Queen Elizabeth II	Feb. 1952–

Capital City: London (1985): 6,767,500.

Ministries

Office of the Prime Minister

Agriculture, Fisheries, and Food	Lord Advocate's Department
Defence	Lord Chancellor's Department
Education and Science	Northern Ireland Office
Employment	Scottish Office
Energy	Trade and Industry
Environment	Transport
Foreign and Commonwealth Office	Treasury
Health and Social Security	Welsh Office
Home Office	

Governments:

	Prime Minister	Parties represented in government and distribution of ministers
July 1945–Feb. 1950	Attlee I (Lab.)	Lab. (all)
Feb. 1950–Oct. 1951	Attlee II (Lab.)	Lab. (all)
Oct. 1951–Apr. 1955	Churchill (Con.)	Con. (all)
Apr. 1955–Jan. 1957	Eden (Con.)	Con. (all)
Jan. 1957–Oct. 1963	Macmillan (Con.)	Con. (all)
Oct. 1963–Oct. 1964	Douglas–Home (Con.)	Con. (all)
Oct. 1964–Apr. 1966	Wilson I (Lab.)	Lab. (all)
Apr. 1966–June 1970	Wilson II (Lab.)	Lab. (all)
June 1970–Mar. 1974	Heath (Con.)	Con. (all)
Mar. 1974–Apr. 1976	Wilson III (Lab.)	Lab. (all)
Apr. 1976–May 1979	Callaghan (Lab.)	Lab. (all)
May 1979–June 1983	Thatcher I (Con.)	Con. (all)
June 1983–June 1987	Thatcher II (Con.)	Con. (all)
June 1987–	Thatcher III (Con.)	Con. (all)

Regions (Counties in Wales):

	Population 1989
England: South East	16 894 000
England: North-West	6 554 000
England: South-West	4 254 000
England: West Midlands	5 165 000
England: East Midlands	3 733 000
England: East Anglia	1 803 000
England: Yorkshire & Humberside	4 892 000
England: North	3 122 000
Scotland: Highland & Islands	253 000
Scotland: Grampian	453 000
Scotland: Tayside	402 000
Scotland: Fife	339 000
Scotland: Lothian	756 000
Scotland: Borders	100 000
Scotland: Central	270 000
Scotland: Strathclyde	2 489 000
Scotland: Dumfries & Galloway	149 000

	Population. 1976
Wales: Clwyd	376 000
Wales: Dyfed	323 000
Wales: Gwent	440 000
Wales: Gwynedd	225 000
Wales: Mid Glamorgan	540 000
Wales: Powys	101 000
Wales: South Glamorgan	389 000
Wales: West Glamorgan	372 000
Northern Ireland: Belfast	363 000
Northern Ireland: West of the Bann	438 000
Northern Ireland: East of the Bann	737 000

Media:
(*a*) Major newspapers:

	Location	Orientation	Daily circulation
Sun	London	Right of centre	4 050 000
Daily Mirror	London	Left of centre	3 139 000
Daily Express	London	Conservative	1 728 000
Daily Mail	London	Right of centre	1 732 000
Daily Telegraph	London	Right of centre	1 131 000
Guardian	London	Left of centre	507 000
The Times	London	Right of centre	467 000
Financial Times	London	Independent	253 000
Independent	London	Independent	414 000

(*b*) Radio and television: *British Broadcasting Corporation* (BBC) formed in 1927 operates under Royal Charter and is financed by licence fees. BBC broadcasts two television channels. *Independent Broadcasting Authority* (IBA) is a commercial concern that provides television service additional to that of BBC. It runs one television channel and it supervises programmes produced by Channel Four.

Economic Interest Organizations:
Employers' organization: Confederation of British Industry (CBI).
 Central trade union peak association: Trades Union Congress (TUC), founded 1868; membership: 9,000,000 (1985).

Central Statistical Office:
Central Statistical Office
HMSO
London

Further Information:

Balsom, D., and Burch, M. (1980), *A Political and Electoral Handbook for Wales*. Aldershot: Gower.

Budge, I., and McKay, D. (1988) (eds.), *The Changing British Political System: Into the 1990s*. London: Longman.

Leys, C. (1989), *Politics in Britain: An Introduction*. London: Verso.

Parry, R. (1988), *Scottish Political Facts*. Edinburgh: T & T Clark.

Rose, R. (1989), *Politics in England*. Boston: Little, Brown.

—— and McAllister, I. (1982), *United Kingdom Facts*. London: Macmillan.

Sked, A., and Cook, C. (1984), *Post-war Britain: A Political History*. Harmondsworth: Penguin.

20. Canada

State Structure: Federal.

Form of State: Constitutional monarchy; acting Governor-General.

Parliament (bicameral): Federal parliament: Senate: 104 seats, until retirement. House of Commons: 195 seats, 5 years.

Electoral System: Plurality formula.

Main Languages: English (61%); French (26%).

Constitutional Development: Important acts passed by the British Parliament are: The Quebec Act of 1774, the Constitutional Act of 1791, the Act of Union of 1840, the British North America Act of 1867, the Canada Act of 1982.

Heads Of State:

King George VI	Dec. 1936–Feb. 1952
Queen Elizabeth II	Feb. 1952 –

Governors–General:

Viscount Alexander of Tunis	Apr. 1946–Feb. 1952
Vincent Massey	Feb. 1952–Sept. 1959
Georges Philias Vanier	Sept. 1959–Mar. 1967
Roland Michener	Apr. 1967–Jan. 1974
Jules Léger	Jan. 1974–Jan. 1979
Edward Richard Schreyer	Jan. 1979–May 1984
Jeanne Sauvé	May 1984 –

Capital city: Ottawa (1987): 301,000.

Ministries

Office of the Prime Minister

Agriculture	National Defence
Communications	National Health and Welfare
Consumer and Corporate Affairs	Public Works
Employment and Agriculture	Regional Industrial Expansion
Energy, Mines, and Resources	Revenue Canada
Environment Canada	Science and Technology
External Affairs	Secretary of State of Canada
Finance	Solicitor-General

Office of the Prime Minister

Fisheries and Oceans
Indian and Northern Affairs
Justice
Labour

Supply and Services
Transport
Treasury Board
Veterans' Affairs

Governments:

	Prime Minister	Parties represented in government and distribution of ministers
June 1945–Nov. 1948	King III (Lib.)	Lib. (all)
Nov. 1948–June 1957	St-Laurent (Lib.)	Lib. (all)
June 1957–Apr. 1963	Diefenbaker (Prog. Con.)	Prog. Con. (all)
Apr. 1963–Apr. 1968	Pearson (Lib.)	Lib. (all)
Apr. 1968–Nov. 1972	Trudeau I (Lib.)	Lib. (all)
Nov. 1972–Aug. 1974	Trudeau II (Lib.)	Lib. (all)
Aug. 1974–June 1979	Trudeau III (Lib.)	Lib. (all)
June 1979–Mar. 1980	Clark (Prog. Con.)	Prog. Con. (all)
Mar. 1980–June 1984	Trudeau IV (Lib.)	Lib. (all)
June 1984–Sept. 1984	Turner (Lib.)	Lib. (all)
Sept. 1984–Nov. 1988	Mulroney (Prog. Con.)	Prog. Con. (all)
Nov. 1988–	Mulroney (Prog. Con.)	Prog. Con. (all)

Provinces:

	Population 1985	Capital
Alberta	2 358 000	Edmonton
British Columbia	2 884 700	Victoria
Manitoba	1 070 600	Winnipeg
New Brunswick	719 600	Fredericton
Newfoundland	580 700	St John's
Nova Scotia	879 800	Halifax
Ontario	9 064 200	Toronto
Prince Edward Island	127 400	Charlottetown
Quebec	6 582 700	Quebec
Saskatchewan	1 017 800	Regina
Territories:		
Northwest territories	51 000	Yellowknife
Yukon territories	23 200	Whitehorse

Media:
(*a*) Major newspapers:

	Location	Orientation	Daily circulation
Toronto Star	Toronto	Liberal	498 000
Globe and Mail	Toronto	Right of centre	320 000
Le Devoir	Quebec	Independent	28 000
La Presse	Quebec	Liberal	210 000
Sun	Vancouver	Centre	237 000
Ottawa Citizen	Ottawa	Liberal	192 000
Toronto Sun	Toronto	Conservative	284 000
Gazette	Montreal	Liberal	198 000

(*b*) Radio and television: *Canadian Broadcasting Corporation* (CBC) is the national, publicly owned, broadcasting service. Many privately owned television and radio stations have affiliation agreements with the CBC.

Economic Interest Organizations:
Employers' organization: Canadian Chamber of Commerce; Canadian Manufacturers' Association.
  Central trade union peak association: Canadian Labour Congress, founded 1956; membership: 2,200,000 (1984).

Central Statistical Office:
Statistics Canada
Ottawa K1A 0T6

Further Information:
Clarke, H. D., et al. (1979), *Political Choice in Canada*. Toronto: McGraw-Hill.
Engelman, F. C., and Schwartz, M. A. (1975), *Canadian Political Parties*. Scarborough: Prentice-Hall.
Hockin, T. A. (1976), *Government in Canada*. Toronto: McGraw-Hill.
Meisel, J. (1973), *Working Papers on Canadian Politics*. Montreal: McGill-Queen's U. P.
Verney, D. V. (1986), *Three Civilizations, Two Cultures, One State: Canada's Political Traditions*. Durham: Duke U. P.

21. USA

State Structure: Federal.

Form of State: Republic.

Parliament (bicameral): Congress: Senate: 100 seats, 6 years: House of representatives: 435 seats, 2 years.

Electoral System: Plurality formula.

Main Languages: English (89%); Spanish (6%).

Constitutional Development: The constitution dates from 1787 and it was adopted in 1789. Twenty-six amendments made to the constitution, by 1991.

Heads of State:

H. S. Truman (Dem.)	Aug. 1945–Jan. 1949
H. S. Truman (Dem.)	Jan. 1949–Jan. 1953
D. D. Eisenhower (Rep.)	Jan. 1953–Jan. 1957
D. D. Eisenhower (Rep.)	Jan. 1957–Jan. 1961
J. F. Kennedy (Dem.)	Jan. 1961–Nov. 1963
L. B. Johnson (Dem.)	Nov. 1963–Jan. 1965
L. B. Johnson (Dem.)	Jan. 1965–Jan. 1969
R. M. Nixon (Rep.)	Jan. 1969–Jan. 1973
R. M. Nixon (Rep.)	Jan. 1973–Aug. 1974
G. R. Ford (Rep.)	Aug. 1974–Jan. 1977
J. E. Carter (Dem.)	Jan. 1977–Jan. 1981
R. Reagan (Rep.)	Jan. 1981–Jan. 1985
R. Reagan (Rep.)	Jan. 1985–Jan. 1989
G. Bush (Rep.)	Jan. 1989–

Capital City: Washington (1984): 622,823.

Ministries

Executive office of the President:

White House Office
Council of Economic Advisers
National Security Council
Administration
Federal Procurement Policy
Management and Budget
Council on Environmental Quality
Policy Development
Science and Technology
 Policy
United States Trade
 Representative

Government Departments:

Agriculture	Interior
Commerce	Justice
Defence	Labor
Education	State
Energy	Transportation
Health and Human Services	Treasury
Housing and Urban Development	

Governments:

	President	Parties represented in government and distribution of ministers
Aug. 1945–Jan. 1949	Truman I (Dem.)	Dem. (all)
Jan. 1949–Jan. 1953	Truman II (Dem.)	Dem. (all)
Jan. 1953–Jan. 1957	Eisenhower I (Rep.)	Rep. (all)
Jan. 1957–Jan. 1961	Eisenhower II (Rep.)	Rep. (all)
Jan. 1961–Nov. 1963	Kennedy (Dem.)	Dem. (all)
Nov. 1963–Jan. 1965	Johnson I (Dem.)	Dem. (all)
Jan. 1965–Jan. 1969	Johnson II (Dem.)	Dem. (all)
Jan. 1969–Jan. 1973	Nixon I (Rep.)	Rep. (all)
Jan. 1973–Aug. 1974	Nixon II (Rep.)	Rep. (all)
Aug. 1974–Jan. 1977	Ford (Rep.)	Rep. (all)
Jan. 1977–Jan. 1981	Carter (Dem.)	Dem. (all)
Jan. 1981–Jan. 1985	Reagan I (Rep.)	Rep. (all)
Jan. 1985–Jan. 1989	Reagan II (Rep.)	Rep. (all)
Jan. 1989–	Bush (Rep.)	Rep. (all)

States:

	Population 1980	Capital
Alabama	3 890 061	Montgomery
Alaska	400 481	Juneau
Arizona	2 717 866	Phoenix
Arkansas	2 285 513	Little Rock
California	23 668 562	Sacramento
Colorado	2 888 834	Denver
Connecticut	3 107 576	Hartford
Delaware	595 225	Dover
Florida	9 739 992	Tallahassee

	Population 1980	Capital
Georgia	5 464 265	Atlanta
Hawaii	965 000	Honolulu
Idaho	943 935	Boise
Illinois	11 418 461	Springfield
Indiana	5 490 179	Indianapolis
Iowa	2 913 387	Des Moines
Kansas	2 363 208	Topeka
Kentucky	3 661 433	Frankfort
Louisiana	4 203 972	Baton Rouge
Maine	1 124 660	Augusta
Maryland	4 216 446	Annapolis
Massachusetts	5 737 037	Boston
Michigan	9 258 344	Lansing
Minnesota	4 077 148	St Paul
Mississippi	2 520 638	Jackson
Missouri	4 917 444	Jefferson City
Montana	786 690	Helena
Nebraska	1 570 006	Lincoln
Nevada	799 184	Carson City
New Hampshire	920 610	Concord
New Jersey	7 364 158	Trenton
New Mexico	1 299 968	Santa Fe
New York	17 557 288	Albany
North Carolina	5 874 429	Raleigh
North Dakota	652 695	Bismarck
Ohio	10 797 419	Columbus
Oklahoma	3 025 266	Oklahoma City
Oregon	2 632 663	Salem
Pennsylvania	11 866 728	Harrisburg
Rhode Island	947 154	Providence
South Carolina	3 119 208	Columbia
South Dakota	690 178	Pierre
Tennessee	4 590 750	Nashville
Texas	14 228 383	Austin
Utah	1 461 037	Salt Lake City
Vermont	511 456	Montpelier
Virginia	5 346 279	Richmond
Washington	4 130 163	Olympia
West Virginia	1 949 644	Charleston

	Population 1980	Capital
Wisconsin	4 705 335	Madison
Wyoming	470 816	Cheyenne

Media:
(*a*) Major newspapers:

	Orientation	Daily circulation
USA Today	Independent	1 300 000
Chicago Tribune	Rignt of centre	1 139 000
New York Times	Left of centre	1 002 000
Washington Post	Independent	1 050 000
Los Angeles Times	Left of centre	1 104 000
Wall Street Journal	Independent	748 000
Christian Science Monitor	Independent	147 000

(*b*) Radio and television: Major commercial networks: *Capital Cities/ABC* (ABC): *Columbia Broadcasting System* (CBS); *National Broadcasting Co.* (NBC).

Economic Interest Organizations:
Employers' organization: US Chamber of Commerce; National Association of Manufacturers; The Business Round Table.

Central trade.union peak association: American Federation of Labor and Congress of Industrial Organizations (AFL-CIO): AFL founded 1881, CIO founded 1938, AFL-CIO merged 1955; membership: 13,500,000 (1985); International Brotherhood of Teamsters, Chauffeurs, Warehousemen and Helpers of America (Teamsters), expelled from AFL-CIO 1957; membership: 2,000,000 (1985).

Central Statistical Office:
Bureau of the Census
US Department of Commerce
Washington, DC 20233

Further Information:
Austin, E. W. (1986), *Political Facts of the United States since 1789*. New York: Columbia U. P.
Barone, M., *et al.* (1972–) (eds.), *The Almanac of American Politics*. Washington, DC: Barone & Co.
Janda, K., *et al.* (1987), *The Challenge of Democracy: Government in America*. Boston: Houghton Mifflin.
McKay, D. (1988), *American Politics and Society*. Rev. edn. Oxford:

Blackwell.

Stanley, H. W. and Niemi, R. G. (1988), *Vital Statistics on American Politics*. Washington, DC: CQ Press.

22. Japan

State Structure: Unitary.

Form of State: Constitutional monarchy.

Parliament (bicameral): Kokkai (Diet): House of councillors: 252 seats, 6 years; House of representatives: 512 seats, 4 years.

Electoral System: Plurality formula; multi-member constituencies.

Main Language: Japanese (99%).

Constitutional Development: The present constitution dates from 1946.

Heads of State:

Emperor Hirohito	Dec. 1926–Jan. 1989
Emperor Akihito	Jan. 1990–

Capital City: Tokyo (1985): 8,353,674.

Ministries:

Office of the Prime Minister

Agriculture, Forestry, and Fisheries	Cabinet Legislation Bureau
Construction	Cabinet Secretariat
Education	Defence
Finance	Economic Planning
Foreign Affairs	Environment
Health and Welfare	Management and Co-ordination
Home Affairs	National Land Agency
International Trade and Industry	Hokkaido Development Agency
Justice	Okinawa Development Agency
Labour	Science and Technology
Posts and Telecommunications	Imperial Household
Transport	

Governments:

	Prime Minister	Parties represented in government and distribution of ministers
May 1946–May 1947	Yoshida I (Lib.)	Lib. (all)
May 1947–Feb. 1948	Katayama (Lib.)	Lib. (all)
Feb. 1948–Oct. 1948	Ashida (Lib.)	Lib. (all)
Oct. 1948–Feb. 1949	Yoshida II (Lib.)	Lib. (all)
Feb. 1949–Oct. 1952	Yoshida III (Lib.)	Lib. (12), Prog. (2)
Oct. 1952–May 1953	Yoshida IV (Lib.)	Lib. (all)
May 1953–Dec. 1954	Yoshida V (Lib.)	Lib.-Yoshida (all)
Dec. 1954–Mar. 1955	Hatoyama I (Lib.)	Lib.-Hatoyama (all)
Mar. 1955–May 1955	Hatoyama II (Lib.)	Lib. (all)
May 1955–Dec. 1956	Hatoyama III (LDP)	LDP (all)
Dec. 1956–Feb. 1957	Ishibashi (LDP)	LDP (all)
Feb. 1957–June 1958	Kishi I (LDP)	LDP (all)
June 1958–July 1960	Kishi II (LDP)	LDP (all)
July 1960–Dec. 1960	Ikeda I (LDP)	LDP (all)
Dec. 1960–July 1962	Ikeda II (LDP)	LDP (all)
July 1962–Dec. 1963	Ikeda III (LDP)	LDP (all)
Dec. 1963–Nov. 1964	Ikeda IV (LDP)	LDP (all)
Nov. 1964–Dec. 1968	Sato I (KDP)	LDP (all)
Dec. 1968–Jan. 1970	Sato II (LDP)	LDP (all)
Jan. 1970–July 1972	Sato III (LDP)	LDP (all)
July 1972–Dec. 1972	Tanaka I (LDP)	LDP (all)
Dec. 1972–Dec. 1974	Tanaka II (LDP)	LDP (all)
Dec. 1974–Sept. 1976	Miki I (LDP)	LDP (all)
Sept. 1976–Dec. 1976	Miki II (LDP)	LDP (all)
Dec. 1976–Dec. 1978	Fukuda (LDP)	LDP (all)
Dec. 1978–Nov. 1979	Ohira I (LDP)	LDP (all)
Nov. 1979–July 1980	Ohira II (LDP)	LDP (all)
July 1980–Nov. 1982	Suzuki (LDP)	LDP (all)
Nov. 1982–Dec. 1983	Nakasone I (LDP)	LDP (all)
Dec. 1983–July 1986	Nakasone II (LDP)	LDP-Nakasone (+NLC)
July 1986–Nov. 1987	Nakasone III (LDP)	LDP (all)
Nov. 1987–May 1989	Takeshita (LDP)	LDP (all)
May 1989–Aug. 1989	Uno (LDP)	LDP (all)
Aug. 1989–	Kaifu (LDP)	LDP (all)

Regions:

	Population 1985	Capital
Hokkaido	5 679 500	Sapporo
Tohoku	9 730 000	Sendai
Kanto	36 786 200	Tokyo
Chubu	20 595 000	Nagoya
Kinki	21 828 000	Osaka
Chugoku	7 748 500	Hiroshima
Shikoku	4 227 400	Matsuyama
Kyushu	13 276 000	Fukuoka
Ryukyu	1 179 000	Naha

Media:
(a) Major newspapers:

	Location	Orientation	Daily circulation
Yomiuri Shimbun	Tokyo	Independent (centre right)	5 430 000
Asahi Shimbun	Tokyo	Centre left	7 600 000
Yomiuri Shimbun	Osaka	Independent (centre right)	2.210.000
Asahi Shimbun	Osaka	Centre left	2 180 000
Mainichi Shimbun	Tokyo	Centre left	4 200 000
Mainichi Shimbun	Osaka	Centre left	1 611 000
Sankei Shimbun	Osaka	Right	1 186 000
Nihon Keizai Shimbun	Tokyo	Independent	1 370 000

(b) Radio and television: *Nippon Hōsō Kyōkai* (NHK) (Japan Broadcasting Corporation) is the non-commercial public broadcasting corporation. National Association of Commercial Broadcasters in Japan (MINPOREN) is the association of the commercial broadcasting corporations.

Economic Interest Organizations:
Employers' organization: Nihon Keieisha Dantai Renmei (NIKKEIREN) (Japan Federation of Employers' Associations).

Trade union central peak associations: As of November 1989, the three major union federations merged to form the Nihon Rōdō Kumiai Sōrengōkai, which has about 8 million members (acronym SHINRENGO — English translation 'General Federation of Japanese Labour Unions'). There is a JCP-orientated federation Zenkoku Rōdō Kumiai

Sōrengo, which claims 1.4 million members (acronym ZENRŌREN—English translation 'National Federation of Labour Unions'). There is a group of left-wing socialists, who used to belong to Sōhyō and refused to join the new federation: Kyōtō Soshiki Zenkoku Rōdō Kumiai Renraku Kyōgikai, with 500,000 members (acronym ZENRŌKYŌ—National Labour Union Federation Co-ordinating Council Joint Struggle Organization).

Central Statistical Office:
Statistics Bureau
Management and Co-ordination
19-1 Wakamatsucho
Shinjuku-ku
Tokyo 162

Further Information:
Baerwald, H. (1986), *Party Politics in Japan*. Boston: Little, Brown.
Hrebenar, R. J. (1986), *The Japanese Party System: From One-Party Rule to Coalition Government*. Boulder, Colo.: Westview Press.
Morishima, M. (1984), *Why has Japan 'Succeeded'?* Cambridge: Cambridge U. P.
Ward, R. E. (1978), *Japan's Political System*. Englewood Cliffs, NJ: Prentice-Hall.

23. Australia

State Structure: Federal.

Form of State: Constitutional monarchy; acting Governor-General.

Parliament (bicameral): Federal parliament: Senate: 76 seats, 6 years; House of representatives: 148 seats, 3 years.

Electoral System: Majority formula with alternative vote; single member constituencies.

Main Language: English (99%).

Constitutional Development: The federal constitution dates from 1900.

Heads of State:

King George VI	Dec. 1936–Feb. 1952
Queen Elizabeth II	Feb. 1952–

Governors-General:

Duke of Gloucester	1945–1947
William McKell	1947–1953
Viscount Slim	1953–1960
Viscount Dunrossil	1960–1961
Viscount De L'Isle	1961–1965
Lord Casey	1965–1969
Paul Hasluck	1969–1974
John Kerr	1974–1977
Zelman Cowen	1977–1982
Ninian Stephen	1982–1989
Bill Hayden	1989–

Capital City: Canberra (1989): 277,100.

Ministries:

Aboriginal Affairs
Administrative Sciences
Arts, Sport, the Environment, Tourism, and Territories
Arts and Territories; Minister Assisting the Prime Minister;
Office of the Prime Minsiter and Minister Assisting the Minister for Immigration, Local Government, and Ethnic Affairs
Attorney-General; and Minister Assisting the Prime

Office of the Prime Minsiter
Minister for Commonwealth/
    State Relations
Community Services and Health
Consumer Affairs; and Minister
    Assisting the Treasurer for Prices
Defence
Defence, Science, and Personnel
Employment, Education, and
    Training
Employment and Education Services
Foreign Affairs and Trade
Finance
Housing and Aged Care
Immigration, Local Government,
    and Ethnic Affairs; and Minister
    Assisting the Prime Minister for
    Multicultural Affairs
Industrial Relations; Minister
    Assisting the Prime Minister
    for Public Service Matters;
    and Minister Assisting the
    Treasurer
Industry, Technology, and
    Commerce
Justice
Land Transport and Shipping
    Support
Local Government, and Minister
    Assisting the Prime Minister
    for the Status of Women
Primary Industries and Energy
Resources
Science, Customs, and Small
    Business
Social Security; and Minister
    Assisting the Prime Minister
    for Social Justice
Telecommunications and
    Aviation Support
Trade Negotiations; Minister
    Assisting the Minister for
    Industry, Technology, and
    Commerce; and Minister
    Assisting the Minister for
    Primary Industries and Energy
Transport and Communications
Treasury
Veterans' Affairs

Governments:

	Prime Minister	Parties represented in government and distribution of ministers
July 1945–Nov. 1946	Chifley I (ALP)	ALP (all)
Nov. 1946–Dec. 1949	Chifley II (ALP)	ALP (all)
Dec. 1949–May 1951	Menzies I (Lib.)	Lib. (15), Nat. (5)
May 1951–Jan. 1956	Menzies II (Lib.)	Lib. (13), Nat. (5)
Jan. 1956–Dec. 1958	Menzies III (Lib.)	Lib. (13), Nat. (5)
Dec. 1958–Dec. 1963	Menzies IV (Lib.)	Lib. (13), Nat. (5)
Dec. 1963–Jan. 1966	Menzies V (Lib.)	Lib. (19), Nat. (5)
Jan. 1966–Dec. 1966	Holt I (Lib.)	Lib. (20), Nat. (4)
Dec. 1966–Dec. 1967	Holt II (Lib.)	Lib. (17), Nat. (4)
Dec. 1967–Jan. 1968	McEwen (Lib.)	Lib. (17), Nat. (4)

Australia

	Prime Minister	Parties represented in government and distribution of ministers
Jan. 1968–Feb. 1968	Gorton I (Lib.)	Lib. (9), Nat. (3)
Feb. 1968–Nov. 1969	Gorton II (Lib.)	Lib. (9), Nat. (3)
Nov. 1969–Mar. 1971	Gorton III (Lib.)	Lib. (9), Nat. (3)
Mar. 1971–Dec. 1972	McMahon (Lib.)	Lib. (10), Nat. (3)
Dec. 1972–Nov. 1975	Whitlam (ALP)	ALP (all)
Nov. 1975–Dec. 1975	Fraser I (Lib.)	Lib. (10), Nat. (4)
Dec. 1975–Dec. 1977	Fraser II (Lib.)	Lib. (17), Nat. (6)
Dec. 1977–Nov. 1980	Fraser III (Lib.)	Lib. (21), Nat. (7)
Nov. 1980–Mar. 1983	Fraser IV (Lib.)	Lib. (20), Nat. (6)
Mar. 1983–Dec. 1984	Hawke I (ALP)	ALP (all)
Dec. 1984–July 1987	Hawke II (ALP)	ALP (all)
July 1987–	Hawke III (ALP)	ALP (all)

States and Territories:

	Population 1989	Capital
New South Wales	5 752 800	Sydney
Victoria	4 303 100	Melbourne
Queensland	2 808 100	Brisbane
South Australia	1 420 400	Adelaide
Western Australia	1 579 800	Perth
Tasmania	449 900	Hobarth
Northern Territory	156 400	Darwin
Australian Capital Territory	277 100	Canberra

Media:
(a) Major newspapers:

	Location	Orientation	Daily circulation
Sun News-Pictorial	Melbourne	Right of centre	564 283
Daily Telegraph	Sydney	Right of centre	278 247
Daily Mirror	Sydney	Right of centre	374 125
Herald	Melbourne	Right of centre	195 223

	Location	Orientation	Daily circulation
Sydney Morning Herald	Sydney	Right of centre	258 056
Age	Melbourne	Left of centre	229 153
Australian		Right of centre	138 497
Canberra Times		Right of centre	45 502
Australian Financial Review		Independent	74 055
West Australian	Perth	Right of centre	249 575

(*b*) Radio and television: *Australian Broadcasting Corporation* (ABC) operates a nationwide non-commercial radio and television service: National Broadcasting Service and National Television and SBS (multi-cultural broadcasting channel). Federation of Australian Radio Broadcasters and Federation of Australian Commercial Television represents privately-owned commercial radio and television stations.

Economic Interest Organizations:
Employers' organization: Confederation of Australian Industries (CAI).
  Central trade union peak association: Australian Council of Trade Unions (ACTU), founded 1927; membership: 2,250,000 (1984).

Central Statistical Office:
Australian Bureau of Statistics
POB 10
Belconnen
Australian Capital Territory 2616

Further Information:
Aitkin, D. (1977), *Stability and Change in Australian Politics*. Canberra: Australian National University Press.
Aitkin, Jinks, and Warhurst (1989), *Australian Political Institutions*. Melbourne: Longman Cheshire.
Hughes, C. A. and Graham, B. D. (1968), *A Handbook of Australian Government and Politics 1890-1964*. Canberra: Australian National U. P.
Jupp, J. (1968), *Australian Party Politics*. 2nd edn. Melbourne: Melbourne U. P.
—— (1982), *Party Politics Australia 1966-81*. Sydney: Allen & Unwin.
Starr, G., *et al.* (1978), *Political Parties in Australia*. Melbourne: Heinemann.

24. New Zealand

State Structure: Unitary.

Form of State: Constitutional monarchy; acting Governor-General.

Parliament (unicameral): House of representatives: 97 seats, 3 years.

Electoral System: Plurality formula.

Main Languages: English (100%); Maori (1.6%).

Constitutional Development: As in the United Kingdom constitutional practice is predominantly an accumulation of convention, precedent, and tradition.

Heads of State:

King George VI	Dec. 1936–Feb. 1952
Queen Elizabeth II	Feb. 1952–

Governors-General:

Lord Freyberg	1946–1952
Lord Norrie	1952–1957
Viscount Cobham	1957–1962
Bernard Fergusson	1962–1967
Arthur Porrit	1967–1972
Denis Blundell	1972–1977
Keith Holyoake	1977–1980
David Beattie	1980–1985
Paul Reeves	1985–

Capital City: Wellington (1986): 325,693.

Ministries

Prime Minister's Department

Agriculture and Fisheries	Energy
Civil Defence	Environment
Commerce	External Relations and Trade
Conservation	Forestry
Consumer Affairs	Health
Customs	Inland Revenue
Defence	Internal Affairs
Education	Justice

Labour
Maori Affairs
Pacific Island Affairs
Police
Scientific and Industrial
  Research
Social Welfare

Prime Minister's Department
Statistics
Survey and Land Information
Tourist and Publicity
Transport
Treasury
Women's Affairs.

Governments:

	Prime Minister	Parties represented in government and distribution of ministers
Apr. 1940–Sept. 1943	Fraser I (LP)	LP (all)
Sept. 1943–Nov. 1946	Fraser II (LP)	LP (all)
Nov. 1946–Dec. 1949	Fraser III (LP)	LP (all)
Dec. 1949–Sept. 1951	Holland I (NP)	NP (all)
Sept. 1951–Nov. 1954	Holland II (NP)	NP (all)
Nov. 1954–Sept. 1957	Holland III (NP)	NP (all)
Sept. 1957–Dec. 1957	Holyoake I (NP)	NP (all)
Dec. 1957–Dec. 1960	Nash (LP)	LP (all)
Dec. 1960–Dec. 1963	Holyoake II (NP)	NP (all)
Dec. 1963–Nov. 1966	Holyoake III (NP)	NP (all)
Nov. 1966–Nov. 1969	Holyoake IV (NP)	NP (all)
Nov. 1969–Feb. 1972	Holyoake V (NP)	NP (all)
Feb. 1972–Dec. 1972	Marshall (NP)	NP (all)
Dec. 1972–Aug. 1974	Kirk (LP)	LP (all)
Aug. 1974–Dec. 1975	Rowling (LP)	LP (all)
Dec. 1975–Dec. 1978	Muldoon I (NP)	NP (all)
Dec. 1978–Dec. 1981	Muldoon II (NP)	NP (all)
Dec. 1981–July 1984	Muldoon III (NP)	NP (all)
July 1984–Aug. 1987	Lange I (LP)	LP (all)
Aug. 1987–Aug. 1989	Lange II (LP)	LP (all)
Aug. 1989 -	Palmer (LP)	LP (all)

Regions:

	Population 1986	Capital (main city)
Northland	126 999	Whangarei
Auckland	887 448	Auckland

	Population 1986	Capital (main city)
Thames Valley	58 665	Thames-Coromandel
Bay of Plenty	187 462	Tauranga
Waikato	228 303	Hamilton
Tongariro	40 793	Taupo
East Cape	53 968	Gisborne
Hawke's Bay	140 709	Napier, Hastings
Taranaki	107 600	New Plymouth
Wanganui	69 439	Wanganui
Manawatu	115 500	Palmerston North
Horowhenua	53 592	Levin
Wellington	325 693	Wellington
Wairarapa	39 608	Masterton
Nelson Bays	69 648	Nelson
Marlborough	38 225	Blenheim
West Coast	34 942	Greymouth
Canterbury	348 712	Christchurch
Aorangi	81 294	Timaru
Clutha-Central Otago	48 771	Oamaru
Coastal-North Otago	137 393	Dunedin
Southland	104 618	Invercargill

Media:

(*a*) Major newspapers:

	Location	Orientation	Daily circulation
New Zealand Herald	Auckland		241 000
Auckland Star	Auckland		111 000
Evening Post	Wellington		85 000
Dominion	Wellington		73 000
Press	Christchurch		85 000

(*b*) Radio and television: *Broadcasting Corporation of New Zealand* (BCNZ) is the public broadcasting corporation. Its operating services are Radio New Zealand (RNZ) and Television New Zealand (TVNZ). Commercial radio has been operating since 1936. Advertising is carried on the two public broadcasting channels and on a third privately owned channel which commenced transmission in 1989. There are also plans for Sky pay-TV.

Economic Interest Organizations:
Employers' organization: New Zealand Employers' Federation.

Central trade union peak association: The New Zealand Council of Trade Unions (formed 1987 from a merger of the Combined State Unions and the former Federation of Labour (founded 1937); membership: 523,374 (1987).

Central Statistical Office:
Department of Statistics
Private Bag
Wellington 1

Further Information:
Jackson, K. (1988), *The Dilemmas of Parliament*. Auckland: Allen & Unwin.
New Zealand Official Yearbook.
Palmer, G. (1988), *Unbridled Power*. 2nd edn. Oxford: Oxford U.P.

Data Archives for the Social Sciences

Austria: WISDOM
Maria-Theresienstrasse 9
A-1090 Wien
Austria

Belgium: Belgian Archives for the Social Sciences (BASS)
Bâtiment SH 2 J. Leclerc
Place Montesquieu 1
B-1348 Louvain-la-Neuve
Belgium

Denmark: Dansk Data Arkiv (DDA)
Campusvej 55
DK-5230 Odense M
Denmark

Finland: No data archive

France: Banque de Données Socio-Politiques (BDSP)
Institut d'Études Politiques
BP 34
F-38401 Saint Martin d'Hères
France

Germany: Zentralarchiv für Empirische Sozialforschung
University of Cologne
Bachemerstrasse 40
D-5000 Köln 41
West Germany

Greece: No data archive

Iceland: No data archive

Ireland: Centre for the Study of Irish Elections
University College
Galway
Ireland

Italy: Archivo Dati e Programmi per le Scienze Sociali (ADPSS)
Istituto Superiore di Sociologica
Via G. Cantoni 4
I-20144 Milano
Italy

Data Archives

Japan:	No data archive
Luxembourg:	No data archive
The Netherlands:	Steinmetz Archive (STAR) Social Science Information and Documentation Centre Herengracht 410–412 NL-1017 BX Amsterdam The Netherlands
Norway:	Norsk Samfunnsvitenskapelig Datatjeneste (NSD) University of Bergen Hans Holmboesgate 22 N-5014 Bergen Norway
Portugal:	No data archive
Spain:	No data archive
Sweden:	Svensk Samhällsvetenskaplig Datatjänst (SSD) Box 5048 S-402 21 Göteborg Sweden
Switzerland:	No data archive
Turkey:	No data archive
United Kingdom:	Economic and Social Research Council Data Archive University of Essex Wivenhoe Park Colchester Essex CO4 3SQ England
Canada:	Leisure Studies Data Bank (LSDB) Department of Recreation University of Waterloo Waterloo, Ontario Canada N2L 3G1
	Machine Readable Archives—Public Archives Canada Public Archives Canada 395 Wellington Ottawa, Ontario Canada K1A 0N3
	Social Science Data Library (SSDL) Department of Sociology Carleton Drive Colonel By Drive Ottawa, Ontario Canada K1S 5B6

	University Data Library (UDL) University of British Columbia Room 206 6356 Agricultural Road Vancouver, British Columbia Canada V6T 1W5
USA:	Data and Program Library Service (DPLS) University of Wisconsin 4452 Social Science Building Madison, Wisconsin 53706 USA
	Inter-University Consortium for Political and Social Research (ICPSR) University of Michigan PO Box 1248 Ann Arbor, Michigan 48106 USA
	National Opinion Research Center (NORC) University of Chicago 6030 South Ellis Avenue Chicago, Illinois 60637 USA
	The Roper Center (RC) University of Connecticut Box U-164 Storrs, Connecticut 06268 USA
	Social Science Data Library (SSDL) Institute for Research in Social Science University of North Carolina Chapel Hill, North Carolina 27514 USA
Japan:	No data archive
Australia:	Social Science Data Archive (SSDA) Australian National University PO Box 4 Canberra ACT 2600 Australia
New Zealand:	No data archive

References

AHLUWALIA, M. S. (1976), 'Inequality, Poverty and Development', *Journal of Development Economics*, 3: 307–42.
BARRET, D. B. (1982) (ed.), *World Christian Encyclopedia*. Oxford: Oxford U.P.
BOLLEN, K. (1980), 'Issues in the Comparative Measurement of Political Democracy', *American Sociological Review*, 45: 370–90.
BORNISCHER, V. (1978), 'Einkommensungleichheit innerhalb von Ländern in komparativer Sicht', *Schweizerischer Zeitschrift für Soziologie*, 4: 3–45.
CASTLES, S. (1984), *Here for Good: Western Europe's New Ethnic Minorities*. London: Pluto Press.
—— and KOSACK, G. (1973), *Immigrant Workers and Class Structure in Western Europe*. Oxford: Oxford U.P.
Encyclopaedia Britannica (1988), *Britannica Book of the Year*. Chicago: Encyclopaedia Britannica.
GASTIL, R. D. (1987) (ed.), *Political Rights and Civil Liberties 1986–1987*. New York: Greenwood Press.
HARTMANN, J. (1984), *Politische Profile der westeuropäischen Industriegesellschaft: Ein vergleichendes Handbuch*. Frankfurt am Main: Campus.
HUMANA, C. (1983), *World Human Rights Guide*. London: Hutchinson.
ILO (1985), *The Cost of Social Expenditure*. Genève: ILO.
—— (various years), *Yearbook of Labour Statistics*. Genève: ILO.
Interparliamentary Union (1986), *Parliaments of the World: A Comparative Reference Compendium*. 2nd edn. Aldershot: Gower.
Keesing's Record of World Events (various years), Harlow: Longman.
KJELLBERG, A. (1983), *Facklig organisering i tolv länder*. Lund: Arkiv.
KORPI, W. (1983), *The Democratic Class Struggle*. London: Routledge.
LIJPHART, J. (1984), *Democracies*, New Haven, Conn.: Yale UP.
MACKIE, T. T. (various years), 'General Elections in Western Nations during [the year in question]', *European Journal of Political Research*.
—— and ROSE, R. (1982), *The International Almanac of Electoral History*. 2nd edn. London: Macmillan.
MADSEN, E. S, and PALDAM, M. (1978), *Economic and Political Data for the Main OECD-Countries 1948–1975*. Århus University: Institute of Economics.
MATHESON, D. K. (1979), *Ideology, Political Action and the Finnish Working Class: A Survey Study of Political Behaviour*. Helsinki: Societas Scientiarium Fennica.
MIELKE, S. (1983) (ed.), *Internationales Gewerkschaftshandbuch*. Opladen: Leske & Budrich.

MITCHELL, B. R. (1981), *European Historical Statistics 1750–1975*. London: Macmillan.
—— (1982), *International Historical Statistics: Africa and Asia*. London: Macmillan.
—— (1983), *International Historical Statistics: The Americas and Australasia*. London: Macmillan.
MULLER, E. N. (1985), 'Income inequality, regime repressiveness, and political violence', *American Sociological Review*, 50: 47–61.
—— (1988), 'Democracy, Economic Development, and Income Inequality', *American Sociological Review*, 53: 50–68.
MUSGRAVE, R. A., and Jarrett, P. (1979), 'International Redistribution', *Kyklos*, 32: 541–58.
NOHLEN, D. (1978), *Wahlsysteme der Welt: Daten und Analysen*. Munich: Piper
Nordic Council (1984), *Yearbook of Nordic Statistics 1983*. Stockholm: Nordic Council.
OECD (1979), *OECD Observer*, 97.
—— (1985a), *Labour Force Statistics 1963–1983*. Paris: OECD.
—— (1985b), *OECD Employment Outlook*. Paris: OECD.
—— (1985c), *Social Expenditure 1960–1990: Problems of Growth and Control*. Paris: OECD.
—— (1985d), *Measuring Health Care 1960–1983: Expenditure, Costs and Performance*. Paris: OECD.
—— (1986), *Living Conditions in OECD Countries: A Compendium of Social Indicators*. Paris: OECD.
—— (1987), *OECD Observer*, 145.
—— (1988), *Labour Force Statistics 1966–1986*. Paris: OECD.
—— (various years), *National Accounts*. Paris: OECD.
—— (various years), *Economic Outlook: Historical Statistics*. Paris: OECD.
—— (various years), *Economic Outlook*. Paris: OECD.
—— (various years), *Revenue Statistics*. Paris: OECD.
PALOHEIMO, H. (1986), *Governments in Democratic Capitalist States 1950–1983: A Data Handbook*. University of Turku, Department of Sociology and Political Science.
PATHIRANE, L., and BLADES, D. W. (1982), 'Defining and Measuring the Public Sector: Some International Comparisons', *Review of Income and Wealth*, 28: 261–89.
PAUKERT, F. (1973), 'Income Distribution at Different Levels of Development: A Survey of Evidence'. *International Labour Review*, 108: 97–125.
RUSTOW, D. A. (1967), *A World of Nations: Problems of Political Modernization*. Washington, DC: The Brookings Institution.
SAWYER, M. (1976), 'Income Distribution in OECD Countries', *OECD Economic Outlook: Occasional Studies*, July: 3–36.
SIPRI (1980), *World Armaments and Disarmament*. London: Taylor & Francis.
—— (1988), *World Armaments and Disarmament*. Oxford: Oxford UP.
SMITH, G. (1984), *Politics in Western Europe: A Comparative Analysis*. 4th edn. London: Heinemann.
The Statesman's Year-Book, 1989–90 (1989). London: Macmillan.

STEPHENS, M. (1976), *Linguistic Minorities in Western Europe*. Llandysul: Gomer Press.

SUMMERS, R., and HESTON, A. (1988), 'A New Set of International Comparisons of Real Product and Price Levels Estimates for 130 Countries, 1950–1985', *Review of Income and Wealth*, 34: 1–25.

TAYLOR, C. L. (1983), *World Handbook of Political and Social Indicators*. 3rd edn. New Haven, Conn.: Yale UP.

—— and HUDSON, M. (1972), *World Handbook of Political and Social Indicators*. 2nd edn. New Haven, Conn.: Yale UP.

TESNIÈRE, L. (1928), 'Statistiques des langues de l'Europe', in A. Meillet, *Les Langues dans l'Europe nouvelle*. Paris: Payot.

THERBORN, G. (1984), 'The Prospects of Labour and the Transformation of Advanced Capitalism', *New Left Review*, 145: 5–38.

United Nations (various years), *Demographic Yearbook*. New York: United Nations.

—— (various years), *Statistical Yearbook*. New York: United Nations.

UUSITALO, H. (1975), *Income and Welfare: A Study of Income as a Component of Welfare in Scandinavian Countries in the 1970's*. Helsinki: Research Group for Comparative Sociology.

VANHANEN, T. (1984), *The Emergence of Democracy: A Comparative Study of 119 States, 1850–1979*. Helsinki: Societas Scientiarum Fennica.

WALLECHINSKY, D., WALLACE, I., and WALLACE, A. (1980), *The Book of Lists*. London: Corgi.

World Bank (1984), *World Tables*. i: *Economic Data*. ii: *Social Data*. 3rd edn. Baltimore, Md.: Johns Hopkins UP.

—— (1988), *World Tables 1987*. 4th edn. Washington, DC: World Bank.

—— (various years), *The World Bank Atlas*. Washington, DC: World Bank.

—— (various years), *World Development Report*. New York: Oxford UP.

Worldmark (1984), *Worldmark Encyclopedia of Nations*. New York: John Wiley